# Practical Agricultural Chemistry

# Practical Agricultural Chemistry

**— 2nd Edition —**

**S.J.M. AULD**

*D.Sc. (Lond.) Ph.D. (Würzburg), F.I.C., F.C.S.*

*Sometime Professor of Agricultural Chemistry at University College, Reading; Late of the South-eastern Agricultural College, and of the Imperial Institute*

**D.R. EDWARDES-KER**

*O.B.E., M.A. (Oxon.), B.Sc. (Lond.)*

*Principal of the Seale-Hayne Agricultural College, Newton Abbot, Devon; Sometime Lecturer in Agricultural Chemistry, and Head of the Chemical Department, South-eastern Agricultural College (University of London), Wye Kent*

BIOTECH

**2007**

**Biotech Books**

**Delhi - 110 035**

ISBN 81-7622-176-7

*Published by* : **BIOTECH BOOKS**
1123/74, Tri Nagar,
DELHI – 110 035
Phone: 27382765
e-mail: biotechbooks@yahoo.co.in

*Showroom* : 4762-63/23, Ansari Road, Darya Ganj,
NEW DELHI - 110 002
Phone: 23245578, 23244987

*Printed at* : **Chawla Offset Printers**
New Delhi – 110 052

**PRINTED IN INDIA**

# PREFACE

THIS book is intended as a practical handbook in Agricultural Chemistry for Students working through courses of instruction for the London B.Sc. degree in Agriculture and other examinations of similar type and standard.

In order to avoid the baldness that cannot be dissociated from a mere list of practical experiments, a short theoretical discussion has been given where necessary before each series of operations, in order to recall to the mind of the student the more salient points in connection with the practical work he has in hand.

Emphasis has been placed on the qualitative side of the subject to a greater extent than is frequently done. It is a matter of extreme regret that many of the students who leave our Agricultural Colleges do so with very little appreciation of the practical value of chemistry to agriculture, and with no more than a theoretical insight into the mechanism and meaning of the countless changes with which they are destined continually to come in contact during their agricultural career. This is often due to their having spent most of their time in the chemical laboratory in working through analytical processes, the intricacies of which have had little meaning for them. It seems to us that the teacher of Agricultural Chemistry should insist on his students using quantitative work largely as a means of interpreting principles or expressing ideas, and not regarding it as an end in itself. It must be borne in mind that those taking such a course of study are primarily agriculturists and not chemists.

Nevertheless, with students of a higher standard, and with those intending to specialise, quantitative Agricultural Chemistry has its place, and the methods given in this book have consequently been carefully chosen, and may be relied upon, as standard or approved processes.

Throughout the book a fair knowledge is assumed on the part of the student of the commoner qualitative and quantitative processes of general chemistry, while in cases of estimations which are not generally included in a course of pure chemistry, such as, for example, the determination of the Iodine Value, Reichert-Meissl number, etc., full practical directions are given. It may be also mentioned that all the experiments described in the text have been personally worked through by one or both of the authors.

The authors are indebted to the standard text-books on the different subjects included under the head of Agricultural Chemistry, especially Hall's *Soil*, Hall and Russell's *Practical Exercises in Agricultural Chemistry*, and Droop Richmond's *Dairy Chemistry;* they also wish to thank Mrs Auld and Mr L. S. Charleton for the line drawings given throughout the book, Mr H. Wormald, B.Sc., for the photomicrographs of starch granules, Messrs R. H. Carter and J. Amos for the photographs of apparatus, and Mr R. H. Carter, in particular, for preparing the index.

S. J. M. A.,
READING.

D. R. E.-K.,
WYE, KENT.

# PREFACE TO SECOND EDITION

IN preparing the second edition, I have not had the advantage of the co-operation of Dr Auld, owing to his absence abroad in an appointment unconnected with Agricultural Chemistry.

The main scheme of the book has been found sufficiently satisfactory to warrant no change whatever in the general arrangement. Without disturbing this, however, several minor alterations have been made in the text, and a few additional experiments included, such as, for example, the perchlorate method for potash estimation.

It is hoped that the book, in this new edition and at the higher price resulting from post war conditions, will still continue to be of value to those students engaged in the study of the scientific side of Agriculture.

My best thanks are due to Mr T. J. Shaw, N.D.A., for the care he has taken in compiling the new index of this edition.

D. R. E.-K.,
NEWTON ABBOT, DEVON.

# CONTENTS

## SECTION I.—PLANT LIFE

### CHAPTER I

#### ULTIMATE CONSTITUENTS OF PLANTS

PAGE

1. Qualitative Detection of Nitrogen . . . 3
2. Qualitative Detection of Phosphorus . . . 4
3. Qualitative Detection of Sulphur . . . 5
4. Quantitative Determination of Moisture . . . 6
5. Quantitative Determination of Ash . . . 7
6. Qualitative Examination of Ash Constituents . . 8
7. Quantitative Estimation of Nitrogen by Kjeldahl's Process . . . . . . 10


### CHAPTER II

#### PROXIMATE CONSTITUENTS OF PLANTS

*(a) Proteins.*

8. Preparation of Gluten from Wheat . . . 14
9. Qualitative Reactions of the Proteins . . . 15
10. Quantitative Determination of Protein . . . 17


*(b) Amides.*

11. Estimation of Amido-nitrogen in Asparagine . . 19


*(c) Fats and Oils.*

12. Quantitative Estimation of Fat or Oil . . . 19
13. Determination of Iodine Value of Linseed Oil . . 22
14. Qualitative Examination of Cotton-seed Oil . . 24

*(d) Carbohydrates.*

PAGE
15. Preparation and Qualitative Examination of Cellulose . 26
16. Qualitative Examination of the Sugars . . . 28
17. Estimation of Sugars by Polarimeter . . . 31
18. Determination of Specific Rotation of Cane Sugar . 32
19. Inversion of Cane Sugar . . . . . 33
20. Preparation and Qualitative Examination of Starch . 35

## CHAPTER III

### PROXIMATE CONSTITUENTS OF PLANTS (*continued*)

*(e) Alkaloids.*

21. Extraction of Caffein from Tea . . . 37
22. Estimation of Nicotine in Tobacco . . . 38
23. Qualitative Reactions of the Alkaloids . . 40

*(f) Glucosides.*

24. Hydrolysis of Salicin . . . . . 42
25. Hydrolysis of Amygdalin. (Cyanogenesis) . . 43

*(g) Organic Acids.*

26. Preparation and Qualitative Examination of Crystallised Oxalic Acid . . . . . . 45
27. Qualitative Reactions of Tartaric Acid and Tartrates . 46
28. Preparation of Citric Acid from Lemons. . . 47
29. Qualitative Reactions of Citric Acid . . . 49

*(h) Essential Oils.*

30. Preparation of Essential Oils by Distillation with Steam 51

*(i) Miscellaneous Substances—Tannins, Resins, Chlorophyll.*

31. Qualitative Examination of Tannin . . . 52
32. Qualitative Determination of Resins in Hops . . 53
33. Preparation and Qualitative Examination of Chlorophyll 54

CHAPTER IV

CHEMICAL CHANGES DURING GERMINATION

PAGE

34. Conversion of Starch to Sugar . . . . 55

*Photo-synthesis.*

35. Plant Catalase . . . . . . 56

SECTION II.—SOILS

CHAPTER V

PROXIMATE CONSTITUENTS OF SOILS

*Limestone, Humus, Sand, and Clay.*

36. Limestone . . . . . . . 57
37. Humus . . . . . . . 59
38. Sand . . . . . . . 60
39. Clay . . . . . . . 61

*Plant Nutrients in the Soil.*

40. Phosphoric Acid, Potash and Nitrates . . . 62

CHAPTER VI

CHEMICAL PROPERTIES OF SOIL

*Nitrification.*

41. Nitrification of Ammonium Compounds . . . 65
42. Effect of Acidity . . . . . . 66
43. General Occurrence of Nitrifying Bacteria . . 66

*Denitrification.*

44. Measurement of Denitrification . . . . 66

*Ammonification.*

45. Measurement of Ammonification . . . . 67

CONTENTS

*Retention of Substances by Soil.*

PAGE

46. Absorption of Colouring Matter and Free Ammonia . 68
47. Absorption of Salts (Qualitative) . . . . 68
48. Absorption of Salts (Quantitative) . . . 69
49. Removal of Lime from the Soil . . . . 71

CHAPTER VII

PHYSICAL PROPERTIES OF SOIL

*Colour of Soil.*

50. Observation of Colour . . . . 73

*Specific Gravity.*

51. Determination of True Specific Gravity . . . 74
52. Determination of Apparent Specific Gravity . . 74

*Pore-space.*

53. Direct Determination in Sand . . . . 74
54. Pore-space of Soils by Calculation . . . 75

*Absorption of Water by Soils.*

55. Hygroscopic Moisture . . . . . 75
56. Retention of Water by Soils . . . . 76
57. Percolation of Water in Soils . . . . 77

*Surface Tension and Capillarity.*

58. Measurement of Capillarity . . . . 78
59. Capillary Attraction in Soils . . . . 79

*Soil Thermometry.*

60. Absorption of Heat by Soil . . . . 81

*Flocculation and Deflocculation of Soil.*

61. Effect of Acids, Alkalies, Salts, etc. . . . 82
62. Shrinkage of Clay on Drying . . . . 83

## CHAPTER VIII

### MECHANICAL ANALYSIS OF SOIL

PAGE
63. Sampling Soil for Analysis . . . . 83
64. Preparation of the Sample . . . . 84
65. Mechanical Analysis by Sedimentation . . . 85
66. Scheme for the Mechanical Analysis of Soil . . 87
67-68. Mechanical Analysis by the Hydraulic Method—Schöne's Elutriator . . . . . 88

## CHAPTER IX

### CHEMICAL ANALYSIS OF SOIL

69. Sampling . . . . . . . 91
70. Moisture . . . . . . . 92
71. Loss on Ignition . . . . . . 92
72. Organic Matter . . . . . . 92
73. Nitrogen . . . . . . . 93

*Total Carbonate.*

74. Determination of Amount of Carbonate . . . 94
75. Method of Absorption . . . . . 94
76. Amos' Method . . . . . . 96

*"Total" Mineral Constituents.*

77. Hydrochloric Acid Extraction . . . . 98
78. Phosphoric Acid . . . . . . 98
79. Potassium . . . . . . . 99
80. Iron and Aluminium . . . . . 100
81. Iron . . . . . . . . 100
82. Manganese . . . . . . . 101
83. Calcium . . . . . . . 101
84. Magnesium . . . . . . . 102
85. Sulphuric Acid . . . . . . 102
86. Nitrates . . . . . . . 102

*"Available" Phosphoric Acid and Potash.*

PAGE

87. Citric Acid Extraction . . . . . 103

*Statement of Results.*

88. Alluvial Soil (Kent) . . . . . 104

## SECTION III.—FERTILISERS AND MANURES

### CHAPTER X

ARTIFICIAL NITROGENOUS MANURES

*Sulphate of Ammonia.*

89. Ammoniacal Liquor . . . . . 105
90. Sulphate of Ammonia . . . . . 106

*Nitrate of Soda.*

91. Qualitative Examination of Nitrate of Soda . . 108
92. Analysis of Nitrate of Soda . . . . . 108
93-95. Estimation of Nitrogen in Sodium or Potassium Nitrate—
  (*a*) Nitrometer Method . . . . . 110
  (*b*) Schlössing's Method . . . . . 112
  (*c*) Estimation by Reduction to Ammonia . 114
96. Presence of Perchlorates in Nitrate of Soda . . 114

*Nitrate of Lime.*

97. Hygroscopicity of Nitrate of Lime . . . 115

*Calcium Cyanamide or Nitrolim.*

98 Qualitative Examination . . . . . 116
99. Analysis of Nitrolim . . . . . 116

CHAPTER XI

ORGANIC NITROGENOUS MANURES

*Farmyard Manure.*

PAGE

100. Examination of Farmyard Manure . . . 117

*Liquid Excrement.*

101. Separation of Urea from Human Urine . . 118
102. Experiments with Urea . . . . . . 119
103. Uric Acid . . . . . . . 120
104. Hippuric Acid . . . . . . 120

*Shoddies, Wool Wastes, etc.*

105. Qualitative Examination of Fibrous Waste Products—
Shoddies, etc. . . . . . . . 121
106. Analysis of Fibrous Waste Products . . 123
107. Examination of other Nitrogenous Waste Products . 123

CHAPTER XII

PHOSPHATIC MANURES

108. The Calcium Phosphates . . . . 125
109. Solubilities of the Calcium Phosphates . . . 125

*Bones and Bone Manures.*

110. Examination of Bones and Bone Compounds . . 127

*Rock Phosphates and Superphosphate.*

111. Examination of Rock Phosphates . . . 129
112. Qualitative Examination of Superphosphate . . 129

*Analysis of Superphosphate of Lime.*

113. Moisture and Combined Water . . . . 130
114. Organic Matter . . . . . . 130
115. Sand, Silica, etc. . . . . . 130

PAGE

116. Phosphates . . . . . . 130
117. The Ammonium Citrate Solution . . . 132
118. Calcium . . . . . . . 133
119. Alkali Salts . . . . . . 133
120. Sulphate . . . . . . . 134
121. Statement of Results . . . . 134

*Basic Slag.*

122. Qualitative Examination of Basic Slag . . . 135
123. Adulteration of Basic Slag . . . . 135
124. Analysis of Basic Slag . . . . . 136

CHAPTER XIII

POTASH MANURES

125. Wood Ashes . . . . . . 137
126. Potash Felspar . . . . . . 137

*Examination of Potash Salts.*

127. Qualitative Examination . . . . 138

*Analysis of Potash Salts.*

128. Potash . . . . . . . 138
129. Chloride . . . . . . . 139
130. Sulphate . . . . . . . 139
131. Lime . . . . . . . 139

CHAPTER XIV

MIXED MANURES AND CALCIUM COMPOUNDS

*Mixed Manures—Guano.*

132. Qualitative Examination of Guanos . . . 140
133. Separation of Uric Acid from Guano . . . 141
134. Examination of other Mixed Manures . . . 142
135. Artificial Manures—Incompatible Manures . . 142
136. Analysis of Mixed Manures . . . . 143

*Lime and Limestone.*

PAGE
137. Limestone . . . . . . . 144
138. Lime . . . . . . . . 145
139. Analysis of Agricultural Limes . . . . . 145

# SECTION IV.—FEEDING STUFFS

## CHAPTER XV

### Composition of Feeding Stuffs

*Analysis of Feeding Stuffs.*

140. Moisture . . . . . . . . 148
141. Ash and Sand . . . . . . . 148
142. Crude Fat or Ether Extract . . . . . 149
143. Nitrogenous Constituents . . . . . 150
144. Crude Fibre . . . . . . . 151
145. Soluble Carbohydrates or Sugars . . . . 152
146. Starch . . . . . . . . 153
147. Pentosans . . . . . . . 153

## CHAPTER XVI

### Concentrated Food-stuffs—Oilcakes, Pulses, Cereals, etc.

*Oilcakes.*

148. Qualitative Examination . . . . . 155
149. Qualitative Examination of Linseed for HCN . . 156
150. Estimation of Prussic Acid in Food-stuffs . . 156
151. Crude Fibre in Cotton-seed Cake . . . . 158
152. Examination of Rape Cake . . . . . 158
153. Lupine Seeds . . . . . . . 159
154. Java Beans, Mauritius Beans, etc. . . . 159

*Cereals.*

PAGE

155. Cereal Starches . . . . . . . 160
156. Examination of Wheat—Crude Gluten . . . 160
Gliadin . . . . . . . 161

*Barley and Malt.*

157. Barley . . . . . . . 161
158. Malt . . . . . . . 162
159. Diastase in Malt Extract . . . . 163
160. "Diastatic Capacity" of Cereals . . . 164

*Rice.*

161. Rice Starch and Rice Gluten . . . . 165
162. Maize . . . . . . . 165
163. Cereal By-Products . . . . . 165

## CHAPTER XVII

### ROOTS, GREEN FODDERS, ETC.

*Roots.*

164. General Analysis of Roots—Swedes, Mangels, Sugar Beet . . . . . . . 166
Determination of Sugar in Sugar Beet . . . 167
165. Expression of the Juice . . . . . 167
166. Sugar in Juice . . . . . . 168
167. Purity of Juice . . . . . . 169
168. Sugar in Beet . . . . . . 169
169. Examination of Ensilage — Detection of Organic Acids, etc. . . . . . . 169

CHAPTER XVIII

SECONDARY FEEDING STUFFS. DIGESTIBILITY DETERMINATIONS

*Molasses and Molasses Feeds.*

PAGE

170. Examination of Molasses . . . . 171
171. Molasses Feeds . . . . . . 172
172. Precipitated Phosphate of Lime . . . . 173
173. Spices, Cattle Powders, etc. . . . . 173

*Digestibility of Feeding Stuffs.*

174. Digestibility of Protein . . . . . 174

SECTION V.—DAIRY PRODUCTS

CHAPTER XIX

MILK

*Sampling.*

175. Qualitative Examination of Milk . . . 176

*The Specific Gravity of Milk.*

176. Determination of Specific Gravity by Density Bottle . 178
177. Determination of Specific Gravity by Westphal's Balance . . . . . . . 179
178. Determination of Specific Gravity by Lactometer . 181

*The Acidity of Milk.*

179. Determination of Acidity . . 182
180. Estimation of Dry Matter or Total Solids . . 183

*Fat.*

181. Estimation of Fat by Adam's Paper-coil Method . 185
182. Estimation of Fat by Gerber's Process . . . 185

*Relation between Specific Gravity, Fat, and Total Solids.*

PAGE

Estimation of Total Protein . . . . . 189

*Nitrogenous Substances.*

183. Determination of Total Nitrogen by Kjeldahl's Process 189

*Estimation of Casein.*

Procedure. . . . . . . 190

*Analysis of Sour Milks.*

184. Weibull's Method . . . . . 191

*Adulteration of Milk.*

185. Change in Density of Milk on Skimming and Diluting —Detection of Added Water—Detection of Removal of Cream—Detection of Simultaneous Removal of Cream and Addition of Water or Skimmed Milk . 191

186. Preservatives—Detection of Boric Acid or Borates—Detection of Formaldehyde—Detection of Hydrogen Peroxide—Detection of Salicylic Acid . . 194

CHAPTER XX

BUTTER

187. Determination of Water . . . . . . 196
188. Determination of Fat . . . . . 197
189. Determination of Total Solids not Fat . . . 198
190. Determination of Common Salt . . . . 198
191. Determination of Protein . . . . 199
192 Determination of Reichert-Meissl Number . . 199

*Saponification of Butter-fat.*

193. Determination of Saponification Number . . 203
194. The Unsaturated Acids in Butter-fat . . . 204
195. Determination of the Iodine Value by Wijs' Method . 204

*The Colour of Butter.*

196. Detection of Artificial Colouring Matter . . 206

## CHAPTER XXI

### CHEESE

PAGE
197. Estimation of Water . . . . . 207
198. Estimation of Fat . . . . . 208

*Nitrogenous Substances.*

199. Determination of Total Nitrogen . . . 209
200. Determination of Total Soluble Nitrogenous Compounds . . . . . . . 209
201. Determination of Soluble Protein Nitrogen . . 209
202. Determination of Ammoniacal Nitrogen . . 209
203. Determination of Nitrogen in form of Amides . . 210
204. Determination of Casein . . . . 210
205. Determination of Ash . . . . . 211
206. Estimation of Sodium Chloride . . . . 211
207. Detection of Phosphoric Acid in Ash Constituents . 211

*Rennet.*

208. Determination of Rennet Strength . . . 212

# SECTION VI.—EXAMINATION OF WATERS AND SOAP

## CHAPTER XXII

### ANALYSIS OF WATER

*Preliminary Examination.*

209. Suspended Matter—Colour—Taste and Smell . . 214

*Total Dissolved Solids.*

210. Estimation of Dissolved Solids . . . . 215

*Organic Matter.*

211. Estimation of Organic Matter—Reagents . . 216
212. Estimation of Ammonia . . . . . 217
213. Interpretation of Results from the Ammonia Estimations 220
214. Oxygen Absorbed, or "Moist Combustion" . . 221
215. Absorption of Oxygen . . . . . 222

*Nitrates and Nitrites.*

PAGE

216. Estimation of Nitrites . . . . . . 223
217. Griess-Ilosvay Method . . . . . . 224
218. Total Nitrates and Nitrites . . . . . 224

*Chlorides.*

219. Determination of Chlorides . . . . . 226

*Acidity in Water.*

220. Total Acidity and Residual Acidity . . . 226

*Estimation of Lead.*

221. Qualitative Determination and Quantitative Estimation . . . . . . . 228

*Estimation of Hardness.*

222. Estimation of Hardness by Soap Solution—Table of Hardness . . . . . . . 229

CHAPTER XXIII

SOFTENING WATER FOR SPRAYS—SOFT SOAPS

223. Use of Washing-soda . . . . . . 233
224. Examination of Soft Soap . . . . . 234

INDEX . . . . . . . . . 235

# LIST OF ILLUSTRATIONS


| FIG. | | PAGE |
|---|---|---|
| 1. | Kjeldahl Digestion Flask | 11 |
| 2. | Kjeldahl Distillation Apparatus without Condenser | 12 |
| 3. | Kieldahl Distillation Apparatus with Condenser | 13 |
| 4. | Soxhlet Extraction Apparatus | 21 |
| 5. | Apparatus for Extraction with Volatile Solvents | 21 |
| 6. | Battery of Extraction Apparatus . *photo, facing* | 22 |
| 7. | Diagrammatic Representation of Polarimeter | 31 |
| 8. | Polarimeter Observation Tube | 32 |
| 9. | Photomicrographs of Starch Granules . *photo, facing* | 34 |
| 10. | Apparatus for Distillation with Steam | 40 |
| 11. | Schrötter's Apparatus for estimating Carbonates | 58 |
| 12. | Simplified Schrötter Apparatus | 59 |
| 13. | Apparatus for measuring Percolation through Soil | 77 |
| 14. | Apparatus for measuring Capillarity in Soils | 80 |
| 15. | Mayer's Modification of Schöne's Elutriator | 89 |
| 16. | Estimation of Carbonates in Soil by Absorption | 94 |
| 17. | Amos's Apparatus for Estimation of Carbonates *photo, facing* | 96 |
| 18. | Lunge Nitrometer | 111 |
| 19. | Schlössing's Apparatus for Estimation of Nitrates | 113 |
| 20. | Apparatus for Drying Substances in Inert Gas | 149 |

FIG. PAGE

21. Specific Gravity Bottle . . . . . 178
22. Pyknometer . . . . . . 179
23. Westphal Balance . . . . . . 180
24. Gerber's Test-bottle . . . . . 186
25. Gerber's Centrifuge . . *photo, facing* 186
26. Apparatus for Gerber's Process . . " " 186
27. Water-bath for Gerber Test-bottles . . . 187
28. Apparatus for determining Reichert-Meissl Value of Fats . . . . . . 201
29. Bottle for Nessler Reagent . . . . 216
30. Distillation Apparatus for Water Analysis . . 218
31. Nessler Glass . . . . . . . 219
32. Apparatus for determining Residual Acidity of Water . 227

# PRACTICAL AGRICULTURAL CHEMISTRY

THE processes involved in agricultural chemistry are naturally more obscure and more complex than those studied in the inorganic and organic departments of the pure science, as so many external factors arise to complicate the chemical changes taking place. Owing to the fact that both plants and animals are living beings, and are affected so largely by individuality, environment, and other external causes, the chemistry connected with their growth, feeding, etc., will naturally not admit of such uniform investigation as is applicable to the other branches of chemistry. The soil also is the seat of chemical changes so obscure, so gradual, and at present so little understood, that the pure chemist may well be appalled at the problems which arise, and the difficulties that are to be met with, in a study of the soil from the point of view of plant nutrition.

The practical methods of investigation employed in agricultural chemistry are, it must be admitted, generally of a very empirical standard, although a knowledge of the more exact methods used in pure chemistry is absolutely essential for a correct understanding of the meaning and underlying reasons of the results obtained.

The need for the recognition of such indefinite factors as palatability in the case of animal food-stuffs, the fact that a large number of the investigations on plants can be often carried out only once a year, the existence of many fermentative changes (malting, humification, nitrification) which seem to lie on the border-line between chemistry and biology, the important part that seasonal and climatic changes play in the case of plants and also of animals, complicate to a large extent the precautions that have to be taken in interpreting results, and often render the practical application of laboratory experiments a matter of extreme difficulty.

But still it cannot be denied that chemistry has proved of incalculable benefit to the scientific agriculturist, and without doubt will in the future play a still more important part in the raising of agriculture to the rank of an exact science.

# SECTION I.—PLANT LIFE

## CHAPTER I

### The Ultimate Constituents of Plants

The ultimate constituents of plants may be defined as the actual chemical elements existing in plants, as distinct from the chemical compounds of which those elements are the component parts.

Of all the seventy odd elements at present known, only twelve or so occur in plants in sufficient amounts to be of importance, and of these twelve elements, three—namely, nitrogen, phosphorus, and potassium—are generally regarded as of especial chemical significance.

The ash of plants contains the non-volatile bodies left after ignition, and consists largely of the carbonates and phosphates of calcium and potassium, although these salts need not have been present as such in the living plant. A knowledge of the elements present in the ash is of importance, as these constituents are all obtained by the plant from the soil.

The water or moisture present in plants should strictly not be considered under the head of ultimate constituents.

#### 1. *Qualitative Detection of Nitrogen.*

(*a*) A gram or so of the substance under observation is well mixed with two to three times its weight of soda-lime and heated to redness in a hard-glass tube. The nitrogenous compounds present are converted into ammonia, the presence of which may be detected at the mouth of the tube by its smell and by its action on litmus. (Will and Varrentrap.)

(*b*) A small quantity of the substance is placed in

a small hard-glass tube with a piece of bright metallic sodium about the size of a pea. On the application of heat a violent reaction results, after which the temperature should be raised to bright redness.

The end of the tube while still hot is dipped into a few cc.s. of water contained in an evaporating basin. The tube will crack, and the charred contents are then boiled with the water for a few moments. After filtering off the glass and carbonaceous particles, a few drops of ferrous sulphate solution are added to the filtrate, which is then acidified with a little strong hydrochloric acid. On the further addition of a small quantity of ferric chloride solution, the production of a precipitate of Prussian blue will be indicative of nitrogen in the substance originally taken. This nitrogen combines with the metallic sodium and some of the carbon under the action of heat to form sodium cyanide, NaCN, which is converted into sodium ferrocyanide by the action of the ferrous sulphate. The addition of the ferric chloride to the acid solution results in the precipitation of ferric ferrocyanide, or Prussian blue.

The nitrogen present in plants and plant products is in combination in the form of amines and amides, proteins, nitrates, salts of ammonium, etc.

## 2. *Qualitative Detection of Phosphorus.*

(*a*) A little of the finely chopped or ground substance is mixed with about its own weight of magnesium powder, and heated strongly in a small hard-glass tube. A slight explosion indicates that the magnesium has decomposed the organic substance, and any phosphorus present will have been converted into magnesium phosphide, $Mg_3P_2$. After cooling, a single drop of water is allowed to run on to the residue, whereby the phosphide is decomposed with the production of phosphine, $PH_3$, readily detected at the mouth of the tube by its characteristic smell.

(*b*) About 2 grams of potassium nitrate are fused in a test-tube, and small quantities of the substance under examination dropped into the fused mass from time to time. Rapid oxidation of the organic substance, accompanied by reduction of the potassium nitrate, takes place, and oxides of nitrogen and carbon dioxide are evolved. Any phosphorus present is oxidised to potassium phosphate, the presence of which is detected by the production of the characteristic yellow precipitate of ammonium phospho-molybdate when the cooled product in the test-tube is dissolved in water, the filtered solution acidified with strong nitric acid, and a solution of ammonium molybdate added. The formation of the yellow precipitate is hastened by the application of heat.

Phosphorus, which occurs in largest amount in seeds, exists in the plant in combination as certain proteins and as lecithin, or choline distearyl-glycero-phosphate, a substance occurring especially in peas, beans, and certain fungi.

### 3. *Qualitative Detection of Sulphur.*

(*a*) Plant products containing sulphur when heated with metallic sodium in an exactly similar way to that described above in **1** (*b*) give sodium sulphide. This is detected in the product after heating, when the boiled and filtered solution treated with a solution of sodium nitroprusside gives a violet coloration.

(*b*) Plant products containing sulphur, when oxidised by fused potassium nitrate according to the method mentioned above in **2** (*b*), give potassium sulphate. This is detected by acidifying the solution of the contents of the test-tube with strong hydrochloric acid, and adding a solution of barium chloride. A white precipitate of barium sulphate results.

Sulphur occurs in plants in the form of certain proteins, and in certain organic sulphocyanides, and sulphides, *e.g.*, oil of mustard, oil of garlic.

The detection of nitrogen, phosphorus, and sulphur, as described above, may be carried out with any plants or plant products convenient at the time, such as wheat grain, mustard seed, grass, potatoes, and from the densities of the precipitates the relative amounts of the elements present in different cases may be roughly compared.

### 4. *Quantitative Determination of Moisture.*

The substance under investigation has first to be obtained in a suitable state of division to allow of ready loss of moisture under the action of heat. Leaves or stems are chopped finely, while samples of potatoes, beets, etc., are cut into cubes of not more than $\frac{1}{4}$ inch edge. A known weight (2 to 3 grams) of the substance is then spread evenly in a porcelain or platinum basin previously heated for half an hour in a steam oven, cooled in a desiccator, and weighed. The basin and its contents are heated for twenty-four hours at 100° C. in the steam oven, cooled in a desiccator, and weighed. In order to ensure that all moisture has been lost, it is advisable to continue the heating for another hour and reweigh, and repeat until no further diminution in weight takes place. From the total loss in weight the percentage of moisture is calculated. The dry matter left should be kept for **5**, below.

This method for the determination of moisture is not wholly satisfactory, as in some cases other volatile substances besides water are lost, and, again, certain substances are sometimes present which undergo gradual oxidation (*e.g.*, linseed oil in flax seed). When the error resulting from such oxidation processes would be appreciable, it is necessary to perform the experiment in an atmosphere of hydrogen or coal gas, although this precaution is seldom required in other cases than that of the type mentioned above.

*Percentages of Moisture in Plant Products.*

In any particular case the percentage of moisture is to a certain extent variable, depending upon soil, climatic conditions, etc. Average figures are, however, as follows :—

| | | | |
|---|---|---|---|
| Wheat (grain) | 10·5 | Timothy grass | 62·0 |
| Rice (grain) | 12·4 | Italian rye grass hay | 8·5 |
| Potatoes (tubers) | 79·0 | Wheat straw | 9·6 |
| Mangolds (roots) | 91·0 | Peas (seeds) | 10·5 |

### 5. *Quantitative Determination of Ash.*

The determination of the percentage of mineral constituents in plants is generally performed on the dry matter obtained after the expulsion of moisture (see 4, above).

About 3 grams of the dry substance are weighed out into a platinum basin which has been heated to redness, cooled in the desiccator, and weighed. The basin and its contents are then ignited over a Bunsen flame for about five minutes, and are finally placed in a muffle furnace. The carbonaceous material gradually burns away, and when no charred particles remain, the basin is withdrawn, placed for a few moments on an asbestos mat, then cooled in a desiccator, and weighed. The percentage of ash constituents in the substance taken is then determined by calculation.

In many cases, however, especially in the case of the seeds of cereals, the ash of which contains a large proportion of salts of the alkali metals, and is readily fusible, it is found to be a matter of some difficulty so to adjust the temperature of the muffle that it is sufficiently high to oxidise the organic matter without fusing the mineral salts present. These salts, if fused, surround particles of carbon, and effectively prevent their oxidation. In cases of this trouble arising, it is best to char the substance before ignition at as low a temperature as possible (faintly visible red heat), and then to extract the major portion of the alkali salts by boiling with water. The black residue and the evaporated extract are separately ignited in the muffle, and their weights added together, to give the value for the total ash.

Suitable substances for the above determination are wheat (grain), different grasses and hays, potatoes, etc., the ash obtained from which may be kept for **6**, below.

The compounds present in the ash of plants are generally not present as such in the plant, but have been produced during ignition from various mineral constituents. For example, the metallic phosphates are produced from complex phosphorus-containing bodies such as lecithin, while the carbonates always present in the ash are formed by the action of heat on such compounds as potassium oxalate, potassium malate, etc. The percentage of ash in a plant varies between certain limits, depending upon climatic conditions, stage of growth, and nature of the soil. In the various portions of the same plant also the percentage varies, being generally less in the seeds than in the leaves or stems.

*Percentages of Ash in different Plant Products.*

| | | | |
|---|---|---|---|
| Rice (grain) | 0·4 | Timothy hay | 4·5 |
| Beans | 3·6 | Wheat straw | 5·3 |
| Wheat (grain) | 1·9 | Potatoes (tubers) | 1·0 |

## *Qualitative Examination of Ash Constituents.*

Plant ash, which contains potassium, sodium, calcium, magnesium, iron, phosphate, carbonate, chloride, sulphate, and silica, can naturally be examined by the ordinary methods of qualitative analysis applicable to a mixture of any metallic salts. In order, however, to avoid the somewhat troublesome separation of phosphoric acid necessary in the case of a complete qualitative analysis, the following simplified method of procedure may be employed.

The ash is dissolved in a small quantity of concentrated hydrochloric acid, and evaporated to dryness on the water-bath, the evolution of carbon dioxide on treatment with acid showing the presence of metallic carbonates. Any silicates present are thus converted into insoluble silica; hence, on dissolving the residue in dilute hydrochloric acid and filtering, the silica is removed. After washing this on the filter paper, the presence of silica in the residue may be confirmed by its insolubility in a fused borax bead.

The filtrate after removal of silica is rendered alkaline with ammonia, boiled, and filtered. All the phosphoric acid, together with all or part of the iron, calcium, and magnesium, is thus obtained on the filter paper, and may be examined for iron, calcium, and phosphates by the usual methods of precipitation as follows:—

*Iron.*—A portion of the precipitate is dissolved in dilute hydrochloric acid, and treated in two portions with potassium ferrocyanide and potassium thiocyanate respectively. A blue precipitate in the first case, and a blood-red coloration in the second, confirm the presence of iron.

*Calcium.*—A small quantity of the moist precipitate is dissolved in dilute acetic acid and ammonium oxalate added. The production of a white precipitate of calcium oxalate, soluble in hydrochloric acid, shows the presence of calcium.

*Phosphoric Acid.*—On dissolving a third portion in concentrated nitric acid and warming with $NH_4MoO_4$ solution, the characteristic canary-yellow precipitate of ammonium phospho-molybdate is produced.

The filtrate from the precipitation with ammonia contains all the sodium, potassium, chloride, and part or all of the magnesium, and sulphate, and is examined for these substances as usual:—

*Calcium.*—$NH_4Cl$ and $(NH_4)_2CO_3$ precipitates the calcium as carbonate, confirmed by dissolving the latter in dilute acetic acid and treating with ammonium oxalate (above).

*Magnesium.*—To the filtrate from the precipitation of calcium carbonate, ammonium phosphate is added, whereby the magnesium is thrown down as phosphate.

*Potassium and Sodium.*—The filtrate from the magnesium precipitate is evaporated to dryness and ignited, in order to expel ammonium salts. The presence of sodium and potassium is confirmed by the usual flame colorations, while the presence of the latter element is also shown by precipitation with platinic chloride or sodium cobaltinitrite.

*Sulphates and Chlorides.*—The presence of sulphate and chloride can be most easily demonstrated in a portion of the original ash by dissolving in hydrochloric acid and precipitating with $BaCl_2$ for the sulphate, and by dissolving in nitric acid and precipitating with $AgNO_3$ for the chloride.

### 7. *Quantitative Estimation of Nitrogen by Kjeldahl's Process.*

This process depends upon the fact that, excepting in the case of a few special compounds, all the nitrogen present in the plant product is converted into ammonium sulphate by digesting the substance with concentrated sulphuric acid at as high a temperature as possible. The ammonia is expelled by boiling the resultant solution with caustic soda and is estimated by absorption in a known volume of standard acid.

From 1 to 4 grams of the substance (the amount taken depends upon the probable nitrogen content) are placed in a Kjeldahl flask (Fig. 1), care being taken that the neck of the flask is dry, to prevent any particles of the substance adhering thereto, and 25 c.c. concentrated sulphuric acid free from nitrogen compounds are added. The flask is then placed in an inclined position and heated gently over a naked flame, in a fume cupboard, excessive frothing (due to the evolution of $CO_2$ and $SO_2$) being avoided.

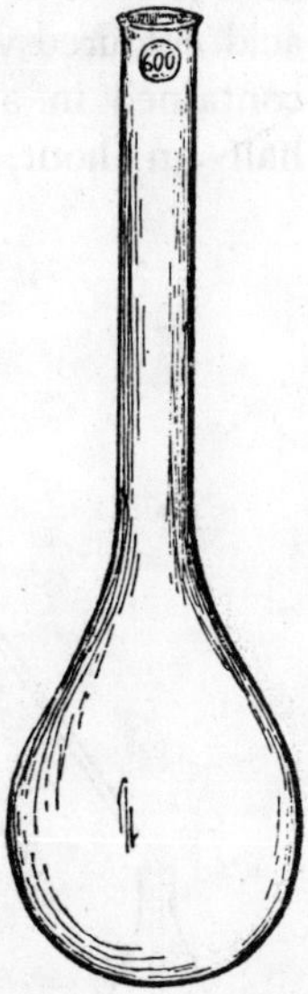

FIG. 1.—Kjeldahl digestion flask.

When the frothing has to a large extent subsided, 10 grams of dry powdered $K_2SO_4$ are added, and the heating is then continued. The object of adding the $K_2SO_4$ is to raise the boiling-point of the acid, and thereby ensure a more rapid oxidation of the carbon present. The liquid which was at first black, gradually becomes lighter in colour and finally quite colourless, at which stage the heating is stopped. If this change takes place only slowly, the addition of a small crystal of copper sulphate will hasten the oxidation.

The contents of the flask are then allowed to cool, and will solidify to a colourless mass of crystals consisting largely of potassium hydrogen sulphate. Sufficient water is added to dissolve the crystals, and the contents of the flask are thoroughly rinsed out into the distilling apparatus shown in Figs. 2 or 3. A few drops of a solution of methyl orange are added and a pinch of zinc dust or a piece of porous tile to ensure even boiling, the cork fitted, and a very concentrated (syrupy) solution of caustic soda run in from the tap funnel with constant shaking until the mixture is decidedly alkaline, as shown by the indicator added. The ammonia liberated is then distilled over into 20 c.c. normal sulphuric acid coloured with a few drops of methyl orange, and contained in a conical flask. After boiling for about half an hour, all the ammonia should have passed

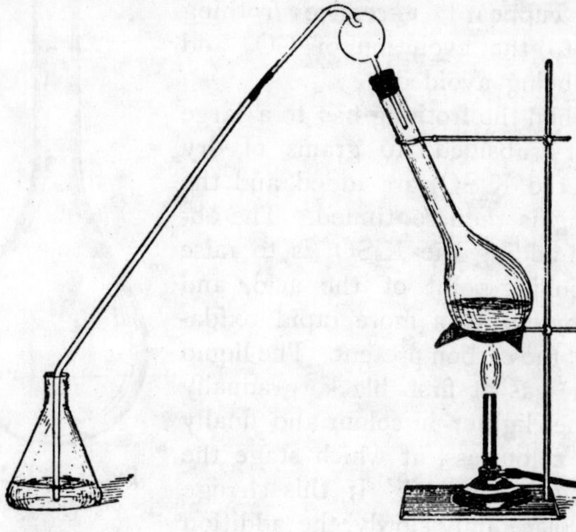

FIG. 2.—Kjeldahl distillation apparatus without condenser.

over, and the end of the delivery tube is then withdrawn from the acid, the adhering liquid washed into the conical flask with distilled water, and the contents of the flask titrated with normal $Na_2CO_3$ or NaOH solution. The amount of normal acid neutralised by the ammonia is thus determined, and from a knowledge that 1 c.c. $\frac{N}{1}$ acid is equivalent to ·014 gram nitrogen in the form of $NH_3$, the percentage of nitrogen in the original substance can be calculated.

The distilling apparatus shown in Fig. 2 avoids the use of a condenser, the steam containing the ammonia delivering directly into the acid. Provided that the solution in the receiving vessel *is kept decidedly acid*, there is no danger of loss of ammonia, even when the temperature has reached the boiling-point.

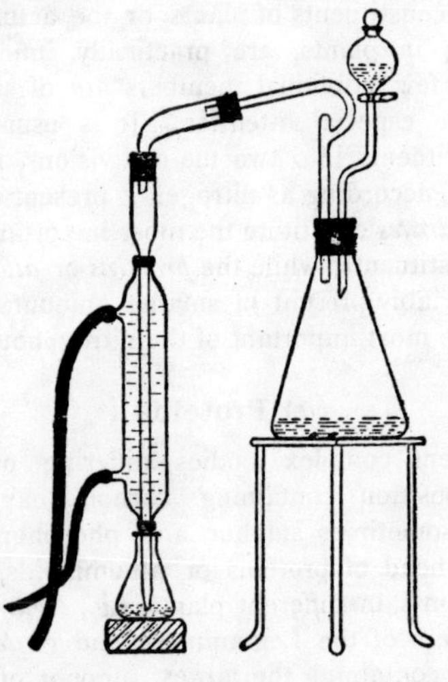

FIG. 3.—Type of Kjeldahl distillation apparatus with condenser.

With the use of a condenser, as in Fig. 3, even in the event of the liquid in the receiver becoming alkaline owing to the presence of insufficient acid, the result of the determination will not generally be rendered valueless.

# CHAPTER II

## Proximate Constituents of Plants

The proximate constituents of plants, or the actual organic compounds existing in plants, are practically infinite in number, although only a few individual members are of sufficient importance to deserve especial attention. It is usual to divide the proximate constituents into two main divisions, nitrogenous and non-nitrogenous, according as nitrogen is present or not.

The *carbohydrates* constitute the most important group of non-nitrogenous constituents, while the *proteins* or *albuminoids*, which are almost invariably present in smaller amount than the carbohydrates, are the most important of the nitrogenous.

### (*a*) **Proteins.**

Many different complex bodies differing only slightly in chemical composition containing carbon, oxygen, hydrogen, nitrogen, and sometimes sulphur and phosphorus, are classed under the one head of proteins or albuminoids, and are found in varying amounts in different plants, *e.g.*, *legumin* or *vegetable casein* in members of the Leguminosæ, and *gluten* in the wheat grain (the latter containing the largest amount of albuminoids of any of the cereals).

### 8. *Preparation of Gluten from Wheat.*

About 50 grams of wheat flour are made into a stiff dough by the addition of sufficient water, and the mass allowed to stand for fifteen minutes or so, in order to allow of the production of gluten from the other proteins by the action of a ferment present. The dough is then placed in a cloth and well kneaded, while a stream of water is allowed to run on to it in order to carry away the starch. After a few minutes the mass may be removed from the cloth, and kneaded directly with the hands in the stream of water until all starch is removed, as shown by the absence of any milky appearance in the water draining from the mass.

The gluten obtained is greyish yellow in colour and elastic, but should not be sticky. The product, however, varies largely, depending upon the quality of the wheat from which the flour was made (see p. 161).

## 9. *Qualitative Reactions of the Proteins.*

Although some reactions are given by all proteins, other tests are responded to by only a certain number. The special colour tests mentioned below will thus be found to succeed only in some cases. Again, some proteins (legumin, vegetable albumins, peptones) are soluble in water, others (globulins) are soluble in saline solutions, while certain insoluble members (gluten) dissolve only in strong acids or alkalies.

The following tests may be carried out with gluten, pea meal (containing legumin), egg albumin, gelatine, and casein, only the first two of which, however, can be regarded as plant products.

*General Reactions* :—

*a.* Heat a little of the substance with dry soda-lime in a hard-glass tube. The protein-nitrogen is converted into ammonia, which can be recognised by smell, turning red litmus blue, etc.

*b. Xantho-proteic Reaction.*—The substance is boiled for some time with concentrated nitric acid. The yellow solution obtained becomes orange on addition of excess of concentrated NaOH solution. This reaction depends upon the formation of nitrophenolic compounds by the oxidising action of the $HNO_3$.

*c. Biuret Reaction.*—To the solution of the substance, if soluble, is added a small quantity of $CuSO_4$ solution and excess of NaOH solution. A violet colour is produced which deepens on heating.

In the case of insoluble proteins, *e.g.* gluten, the substance is dissolved in concentrated NaOH solution with application of heat, and the $CuSO_4$ solution then added.

*Special Reactions* :—

*d. Millon's Reaction.*—Egg albumin and other proteins containing tyrosine or oxyphenyl groups give a red precipitate on boiling with Millon's reagent.[1] In the absence of tyrosine or other bodies containing the oxyphenyl group, this reaction does not take place, and is therefore not obtained in the case of gluten or gelatine.

*e. Adamkiewicz's Reaction.*—Gluten, egg albumin, and other proteins containing tryptophane (indole-$\beta$-amino-propionic acid), on dissolving in glacial acetic acid, and pouring a small quantity of concentrated $H_2SO_4$ down the side of the test-tube, give an intense purple ring at the junction of the two liquids. Hopkins and Cole have shown this reaction to be due to the presence in the acetic acid of glyoxylic acid as an impurity, and the presence of this latter compound should be ensured by adding to the acetic acid, previous to making the test, a small quantity of a solution of oxalic acid that has been partially reduced to glyoxylic acid by sodium amalgam. The glyoxylic acid condenses with the indole-amino-propionic acid formed by the hydrolysis of the albumin, with production of a purple body of complex composition. This reaction is not given by gelatine and other proteins that do not contain tryptophane.

*f. Vogel's Lead Sulphide Reaction.*—This reaction is given only by cystine and proteins containing sulphur, and therefore is not obtained with gelatine.

The substance is boiled for a short time with strong NaOH solution, whereby the molecule is broken up

[1] Millon's reagent should be freshly prepared by dissolving mercury in an equal weight of concentrated nitric acid with application of heat, and then diluting the solution of mercuric nitrate with an equal volume of water.

and sodium sulphide produced. On adding lead acetate solution to the mixture, a black precipitate of lead sulphide is obtained, and from the depth of coloration a rough indication of the amount of sulphur present in the body under investigation is given.

## 10. *Quantitative Determination of Protein.*

The conventional method for the estimation of the percentage of proteins present in an organic substance containing them is to determine the percentage of total nitrogen by Kjeldahl's process, and from an assumption that all this nitrogen exists in the substance in the form of proteins, to calculate the amount of the latter bodies from a knowledge of the proportion of nitrogen contained by them.

This method is, however, open to several objections:—The percentage of nitrogen in different albuminoid substances varies from 14·0 in the case of mucin, to 18·4 in various proteins present in wheat. It is usual, however, to take the mean (16 per cent.) in the calculation of "crude" protein (see p. 150) from the nitrogen percentage, although it can be seen that in many cases an appreciable error will be thus introduced.

Again, various bodies other than proteins, such as amines, ammonium salts, and other nitrogen-containing substances,[1] suffer decomposition during heating with concentrated $H_2SO_4$ according to Kjeldahl's process. It will be seen that the nitrogen present in all those bodies undergoing hydrolysis in this way is given in the analysis under the head of crude protein, although these nitrogenous substances may possess a very different feeding value, or no feeding value at all (for differentiation between different nitrogenous compounds in the case of cheese, see p. 209).

In spite of these obvious objections, however, the method is still largely adhered to in agricultural analysis.

About 4 grams of ground wheat are heated in a Kjeldahl flask with concentrated $H_2SO_4$ in an exactly

[1] Nitrogen in the form of nitrates can also be estimated together with nitrogen in other forms, if salicylic acid be added to the mixture in the Kjeldahl flask, owing to the formation of nitrosalicylic acid, which readily undergoes reduction.

similar manner to that described in **7**, p. 10. The ammonia, after liberation with caustic soda, is distilled off as usual, absorbed by standard acid, and estimated by titration. The percentage of nitrogen so found is multiplied by $5{\cdot}8\ (=\frac{100}{17{\cdot}3})$, 17·3 per cent. being taken as the average percentage of nitrogen in the various proteins in wheat, and the result given under the head of protein.

In the estimation of protein in other substances, different factors are employed, as follows:—

| | |
|---|---|
| Oat proteins . . . . | 6·31 |
| Maize proteins . . . . | 6·22 |
| Flax seed proteins . . . | 5·5 |

An improved method, whereby a differentiation between the true albuminoid nitrogen and nitrogen existing in other forms is made, is being used to a certain extent in agricultural analysis, especially in America. For practical details of the process, see the section on Feeding-Stuffs, p. 150.

### (*b*) Amides.

Amides may be regarded as organic acids in which the hydroxyl of the —COOH group is replaced by the amido group $NH_2$. The amido-radicle is therefore $—CONH_2$, and is almost invariably present, although in different proportions, in proteins and albuminoid bodies, especially the polypeptides; as, for example, $CONH_2.NH.CH_2.CO.NH.CH_2.CONH_2$, one of the simplest compounds of this class.

One of the best methods of estimating nitrogen in the amido form depends upon the fact that the $—CONH_2$ group, on hydrolysis with dilute acids, gives —COOH and the ammonium salt of the acid. The ammonia can be readily liberated from this salt by the action of alkalies, and can be estimated by distillation in the usual way.

Although amido-nitrogen is included under the head of protein nitrogen in ordinary agricultural analysis, the need of distinguishing between nitrogen combined in these two forms is evident when their very great difference in feeding value is remembered.

### 11. *Estimation of Amido-nitrogen in Asparagine.* ($NH_2.CO.CH_2.CH(NH_2).COOH.$)

One gram of asparagine is boiled with 20 c.c. of a 5 per cent. solution of hydrochloric acid for one and a half hours with reflux condenser, the solution cooled, almost neutralised with $Na_2CO_3$, and transferred to an ammonia-distilling apparatus (see Figs. 2 and 3, pp. 12, 13). Three or 4 grams of magnesia are then added directly to the liquid in the flask, and the ammonia distilled over into 30 c.c. $\frac{N}{2}$ $H_2SO_4$ diluted with about an equal volume of water.

After half an hour the receiving vessel is removed, the end of the delivery tube washed into it, and the excess of acid determined by titration with $\frac{N}{2}$ alkali. The number of cubic centimetres of $\frac{N}{2}$ acid neutralised by the ammonia being obtained, the percentage of amido-nitrogen in asparagine can be calculated.

Asparagine contains half its nitrogen in the amido form, and the percentage of amido-nitrogen is therefore $\frac{14 \times 100}{132} = 10{\cdot}6.$

$$NH_2.CO.CH_2.CH(NH_2)COOH + H_2O = COOH.CH_2.CH(NH_2).COOH + NH_3.$$

Aspartic acid.

## (*c*) **Fats and Oils.**

### 12. *Quantitative Estimation of Fat or Oil.*

The natural fats and oils (excluding the essential oils) are the glycerides of the higher saturated or unsaturated fatty acids. Owing to their high feeding value as producers of heat and energy, the determination of the percentages of these compounds in plant products is of the greatest importance (see Ether Extract, **142**, p. 149).

The proportions of fats and oils in different plants vary considerably, being in the case of ordinary crops generally less than 5 per cent., and often less than 1 per cent. Some plants, however, such as flax and cotton, contain in their seeds such a high percentage that they are cultivated very largely for the sake of the oil.

*Percentages of Fats in different Plant Products.*

| | | | |
|---|---|---|---|
| Wheat grain | 2·1 | Potatoes | 0·1 |
| Oats (grain) | 5·0 | Lucerne hay | 2·2 |
| Linseed (flax seed) | 33·7 | Wheat straw | 1·3 |
| Cotton seed | 20·0 | Timothy grass | 1·2 |

In agricultural analysis the fats are generally given under the head of "Ether Extract," or "Crude Fat," together with other substances soluble in ether (see **142**, p. 149).

For the determination of the oil in cotton seed or other oil seed, the sample should be in the dry state, having been previously heated to 100° C. for twelve hours in the steam oven. In the case of linseed and other substances containing unsaturated compounds, this heating must be carried out in an atmosphere of hydrogen or coal-gas. About 3 or 4 grams of the dry sample are accurately weighed out and placed in a filter-paper cartridge which has previously been extracted with ether for some time in a Soxhlet extractor. The cartridge is then placed in a Soxhlet extractor fitted with a condenser and a *weighed* flask (see Figs. 4, 5 and 6). Anhydrous alcohol-free ether is then poured through the condenser on to the cartridge in sufficient amount to start the syphon, and then another 20 c.c. or so are added. The flask is then heated on the water-bath for from two to six hours, depending upon the nature of the sample. The lower the percentage of fat in the latter the longer should be the time of extraction, as in this case any small quantity of unextracted fat, negligible in the case of a substance such as linseed, might lead to a large *percentage* error.

When the extraction is complete, the burners are turned out, the cartridge and its contents removed,

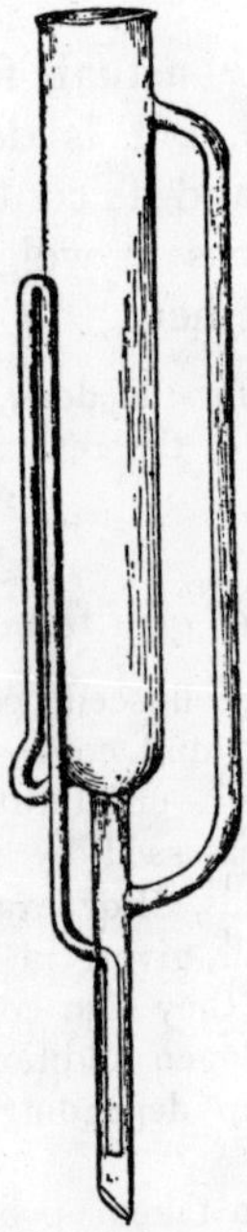

FIG. 4.—Soxhlet extraction apparatus.

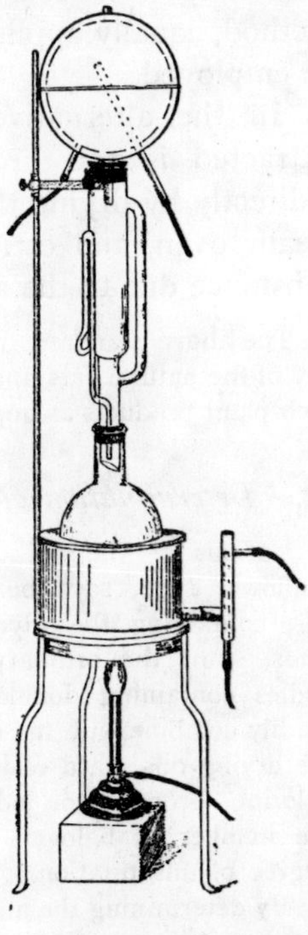

FIG. 5.—Apparatus for extraction with volatile solvents.

the extractor and condenser again fitted, and the latter heated on the water-bath until all the ether has been distilled over into the upper part of the apparatus.

The last traces of ether are driven off by placing the flask for a short time in a steam oven. The flask is then cooled and weighed, and from the increase in weight the percentage of fat in the sample determined. In the case of linseed oil, however, the final heating in the steam oven must be carried out in an atmosphere of hydrogen or coal-gas, or the following alternative method, equally applicable to any substance taken, may be employed.

In the alternative method the actual fat or oil extracted is not directly weighed, but is determined indirectly by drying the cartridge and its contents in a steam oven, and estimating the loss in weight of the substance due to the action of the ether.

The above standard method is used for the determination of any of the natural fats and oils, and also of the resins present in such plant products as hops (see p. 53).

## 13. *Determination of Iodine Value of Linseed Oil.*

Various of the natural oils, notably linseed, poppy seed, sunflower seed, soya bean, cotton seed, and colza (rape seed) oils, containing the glycerides of various unsaturated organic acids, show the ordinary properties possessed by unsaturated bodies containing double or triple bonds. For example, they readily combine with the oxygen of the air, giving in the case of the drying oils, hard resinous products; they also combine with chlorine, bromine, or iodine to give halogen addition products, the number of halogen atoms taken up depending upon the degree of unsaturation.

By determining the amount of halogen taken up by a certain quantity of a given oil, the identity of the oil, and also the presence or absence of impurities, can be established, from a knowledge of the amount of halogen absorbed by the oil in the pure condition.

Hübl's method for the determination of the "iodine value" of natural fats, oils, and waxes is based upon the above consideration, and the estimation of this number is of the utmost importance in the technology of oils.

The "iodine value" is the number of grams of iodine absorbed by 100 grams of the oil in question, and it is found that the value

FIG. 6.—Battery of extraction apparatus, showing different types of condensers.

[Face p. 22.

actually obtained in different cases agrees fairly closely with the theoretical value calculated from the number of double and triple bonds in the molecule.

For example, in the case of oleic acid, $CH_3[CH_2]_7 . CH : CH [CH_2]_7 . COOH$, 2 gram atoms of iodine ($2 \times 127$ grams) should be theoretically required to saturate 1 gram molecule (282 grams) of oleic acid. This gives 90·07 for the number of grams of iodine per 100 grams oleic acid, while the iodine value actually found is 89·8 to 90·5.

| *Organic acid.* | *Double or triple bonds.* | *Atoms of iodine, per molecule acid.* |
|---|---|---|
| Oleic and Ricinoleic Series | One double bond | Two |
| Linoleic | One triple bond | Four |
| Linolenic | Three double bonds | Six |

Hübl's method for the determination of the iodine value of unsaturated acids or their glycerides consists in adding to a known weight of the fat dissolved in chloroform an excess of an alcoholic solution of mercuric chloride containing a known amount of free iodine per cubic centimetre.

The excess of iodine is then determined by titration with standard thiosulphate solution. It is found that the presence of the mercuric salt is essential for the reaction to proceed in a regular, well-defined manner, although the actual rôle played by this compound is not known.

*The Solution of Iodine and Mercuric Chloride* is prepared as follows:—Twenty-five grams iodine are dissolved in 500 c.c. 95 per cent. alcohol, and then added to a solution of 30 grams mercuric chloride, $HgCl_2$, in an equal quantity of the same spirit. The amount of free iodine in the solution is found to vary constantly, and should therefore be standardised by titration with thiosulphate immediately before use.

The chloroform employed for the solution of the fat should be previously tested by treating 10 c.c. with an equal volume of the iodine solution, and titrating the free iodine after two hours standing. The amount of free iodine should be unchanged.

From ·15 to ·2 gram of linseed oil is accurately weighed out, dissolved in 12 c.c. of pure chloroform, and placed in a well-stoppered flask of 500 c.c. capacity. From 25 to 30 c.c. of the iodine solution are then delivered from a pipette into the chloroform solution: the resulting mixture should give a clear solution on shaking. If the solution be not clear, a further small quantity of the chloroform should be added. After standing for two hours, the solution should still possess a deep brown colour, indicative of the presence of excess of free iodine. If the colour has been discharged, an additional 2 or 3 c.c. of the iodine solution should be run in, and the mixture allowed to stand for another two hours. Twenty c.c. of a 10 per cent. solution of KI are then added, and about 300 c.c. distilled water. The excess of iodine is then estimated by direct titration with decinormal sodium thiosulphate solution, with constant agitation. When the colour is almost discharged, a few drops of starch solution are added, and the titration carried to an end.

For an improved method for the determination of the "Iodine Value," especially of butter fat, see Wij's method, p. 204.

The iodine values of the more important oils are as follows:—

| | | | |
|---|---|---|---|
| Linseed oil | 172 | Soya-bean oil | 122 |
| Poppy-seed oil | 134 | Cotton-seed oil | 104 |
| Olive oil | 81 | Colza oil | 100 |

## 14. *Qualitative Examination of Cotton-seed Oil.*

Cotton-seed oil, obtained from the seeds of the various kinds of cotton tree, is of a reddish or greenish brown colour when crude, but after purification is pale yellow. Being inexpensive and of a pleasant taste, it is employed largely for the adulteration of olive oil (salad oil), lard, margarine, etc., and for this reason the working out of methods for the rapid detection of the presence of cotton-seed oil by various colour tests has received a deal of attention.

The following tests should be applied to (i.) pure cotton-seed oil; (ii.) pure olive oil, and (iii.) the latter containing 5, 25, and 50 per cent. of added cotton-seed oil.

(*a*) *Lewkowitsch's Nitric Acid Test.*—One c.c. of the oil is vigorously shaken up in a test-tube with an equal volume of nitric acid of specific gravity 1·37. Pure cotton-seed oil gives a dark brown coloration after being allowed to separate for fifteen minutes or longer, while olive oil containing cotton-seed oil as impurity will give lighter colours; from the intensity of the coloration some idea as to the amount of cotton-seed oil present may be arrived at.

(*b*) *Becchi's Test.*—Ten c.c. of oil are shaken in a test-tube with 1 c.c. of an alcoholic solution of silver nitrate to which a few drops of ether have been added. Ten c.c. of amyl alcohol containing 15 per cent. of rape oil (colza oil) are then added, the mixture well shaken, divided into two equal parts, and one part heated on the water-bath for fifteen minutes.

The production of a dark brown colour is characteristic of cotton-seed oil, although this test does not always give reliable results.

(*c*) *Halphen's Test.*—To 1 c.c. of the oil dissolved in 1 c.c. amyl alcohol in a test-tube are added 1 c.c. of carbon-bisulphide in which 1 per cent. of ground roll sulphur has been dissolved.[1] The mixture is well shaken, then heated on the water-bath for fifteen to twenty minutes. A deep red colour is produced with pure cotton-seed oil, while smaller proportions of the oil in olive oil give lighter colorations.

This colour test is an extremely delicate one, and is, moreover, very reliable, it failing only in the case of cotton-seed oil previously heated to 200° C. The test is sufficiently sensitive to detect cotton-seed oil in the milk of cows fed on cotton-seed cake, and is perhaps the best colour test that can be employed

[1] Flowers of sulphur should not be used, as this contains an allotropic modification of sulphur which is insoluble in $CS_2$.

for the detection of cotton-seed oil in cheap olive oil, margarine, etc.

### (*d*) **Carbohydrates.**

When carbon is assimilated from the $CO_2$ of the air by the action of chlorophyll in conjunction with protoplasm (see p. 54), it is elaborated into sugars, starches, and cellulose. There is still some doubt as to which of these substances is the first product of the assimilation. Brown and Morris (1893) concluded that cane sugar was the primary substance, while according to the later work of Campbell (1912) it was concluded that dextrose and lævulose are the first carbohydrates to be found.

As to the states in which the above carbohydrates exist in plants, the sugars are in solution in the cell sap, the starch is in the solid form of granules in the plant cells, while the cellulose largely makes up the framework of the cells.

### **15.** *Preparation and Qualitative Examination of Cellulose.*

About 2 grams of finely ground straw or hay, or other suitable material, are placed in a flask and boiled for twenty minutes with a mixture of 200 c.c. water and 2 c.c. concentrated sulphuric acid. Various mineral constituents, together with alkaline bodies, etc., are thereby removed. The acid liquid is poured off, the residue thoroughly washed free from acid, and 200 c.c. water containing 2·5 g. caustic soda added. The mixture is again boiled for twenty minutes, the alkaline liquid poured off, and the residue carefully washed as before. The alkali removes the acid constituents, phenols, etc., that may be present. The fibrous product is fairly pure cellulose, but is generally dirty grey in colour, in which case it should be bleached by warming with 200 c.c. of a 50 per cent. solution of HCl and adding a few crystals of potassium chlorate. The nascent chlorine evolved bleaches the colouring matter, and the cellulose is then well washed several times with boiling water.

The following tests should be performed with the product, and also with ordinary filter paper, which is a fairly pure form of cellulose.

(i.) The substance in small amount is well shaken in a corked test-tube with Schweizer's reagent (an ammoniacal solution of copper oxide). The cellulose will be found to be gradually disintegrated, and finally completely dissolved. On pouring the blue solution into a large volume of water, the cellulose will be precipitated as a colloidal hydrate.

(ii.) A similar series of operations is carried out with a concentrated solution of $ZnCl_2$ containing free HCl. The cellulose will be found to dissolve or swell up, but on pouring the product into water, the milky precipitate of zinc oxychloride produced will mask the cellulose.

(iii.) Successive portions of the cellulose are well shaken in an open test-tube with $\frac{1}{2}$ c.c. concentrated $H_2SO_4$. The cellulose will gradually dissolve, being converted into cellulose sulphates. The test-tube is half-filled with water and the solution well boiled for a few minutes, whereby the cellulose sulphates are hydrolysed to glucose. The presence of this latter substance in the solution can be detected by almost neutralising with a concentrated (syrupy) solution of NaOH, and then adding Fehling's solution (alkaline copper tartrate). On boiling, a red precipitate of cuprous oxide is produced, indicative of the presence of glucose (see **16** (*b*), p. 28).

Pure cellulose is a colourless, insoluble material, differing in texture according to its source, which can therefore often be decided from a microscopic examination.

When precipitated from its copper solution (i. above) by dilute acids, it is however, structureless, and dries to a horny, amorphous mass. The digestibility of cellulose in plant products used for food depends largely upon the degree of hydration. In young plants, for example, the cellulose is entirely digestible, but as the plant gets older the digestibility decreases until finally the cellulose in old plants and hard wood (ligno-cellulose) is of no value as a feeding-stuff.

## 16. *Qualitative Examination of the Sugars.*

The natural sugars are divided mainly into two groups, the *monoses*, possessing the general formula $C_6H_{12}O_6$, and the *bioses* of the formula $C_{12}H_{22}O_{11}$, formed by condensation of two monose molecules, with elimination of one molecule of water. The bioses can be split up into monoses by the addition of a molecule of water (hydrolysis) brought about by the action of enzymes, such as diastase (in malt), ptyalin (in saliva), or by boiling with dilute acids.

The monoses (glucose, fructose, etc.) occur in largest amount in ripe fruits, while the best-known biose (cane sugar or sucrose), although also present in fruits, exists in larger proportion in the stalks of certain plants such as sugar cane, sorghum cane, and in various roots, especially in those of the sugar beet, the proportion of sugar in certain samples of the latter reaching 25 per cent.

The following tests should be applied to cane sugar, glucose, and fructose, and also when possible to certain natural sugar-containing substances such as honey, and the aqueous extracts prepared from raisins, sugar beet, and carrots.

(*a*) *Reaction with* $H_2SO_4$.—A small quantity of the solid sugar or its syrupy solution in water is heated with concentrated sulphuric acid. Cane sugar is charred, while glucose is not blackened if pure.

(*b*) *Reaction with Fehling's Solution.*—Reducing sugars on boiling with Fehling's solution (an alkaline solution of cupric tartrate)[1] give a red precipitate of

[1] As Fehling's solution deteriorates on keeping, it is advisable to make up the copper solution and the tartrate solution separately, and only mix immediately before using. With the following proportions, the solution can be used for the quantitative estimation of glucose :—*Solution A.* 17·32 grams $CuSO_4,5H_2O$ dissolved in 150 c.c. water and the cold solution made up to 250 c.c. *Solution B.* 35 grams NaOH and 90 grams sodium potassium tartrate (Rochelle salt) dissolved in 150 c.c. water and the cold solution made up to 250 c.c.

cuprous oxide $Cu_2O$. This reaction is quantitative and can be used for the estimation of sugars (see p. 152).

Cane sugar does not give this reaction unless it has been previously inverted by boiling for a few seconds with a dilute acid, whereby it is converted into equal molecules of glucose and fructose—

$$\underset{\text{Cane sugar}}{C_{12}H_{22}O_{11}} + H_2O = \underset{\text{Glucose}}{C_6H_{12}O_6} + \underset{\text{Fructose.}}{C_6H_{12}O_6}.$$

The sugars on oxidation with Fehling's solution give many different substances, such as tartronic, formic, and oxalic acids.

(*c*) *Reaction with Phenylhydrazine.*—Two grams of glucose or fructose are dissolved in 15 c.c. water and the solution added to 2 grams phenylhydrazine dissolved in just sufficient dilute acetic acid. The mixture is heated for half an hour on the water-bath, when the whole of the sugar is precipitated in the form of its *glucosazone.* Cane sugar does not give this reaction unless inverted, and the acid used for hydrolysis neutralised.

The formation of glucosazone from glucose takes place in three stages with the intermediate formation of glucose phenylhydrazone, thus—

(i.)

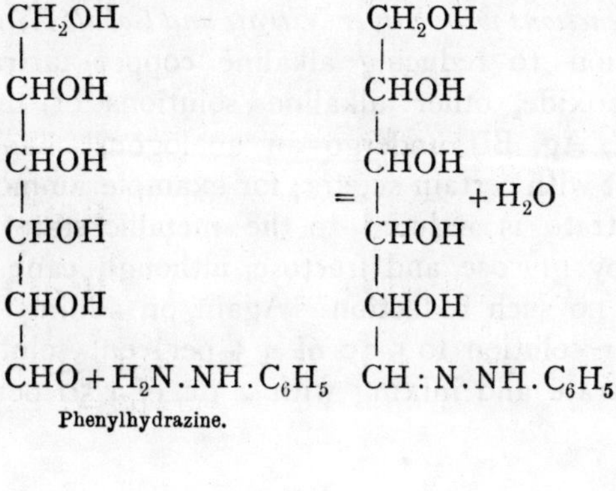

$$\text{(ii.)}\quad \begin{array}{l} CH_2OH \\ | \\ CHOH \\ | \\ CHOH \\ | \\ CHOH \\ | \\ CHOH + H_2N.NHC_6H_5 \\ | \\ CH:N.NHC_6H_5 \end{array} \quad \begin{array}{l} CH_2OH \\ | \\ CHOH \\ | \\ =CHOH + NH_3 + C_6H_5NH_2 \\ | \\ CHOH \\ | \\ CO \\ | \\ CH:N.NHC_6H_5 \\ \text{Hypothetical oxyhydrazone.} \end{array}$$

$$\text{(iii.)}\quad \begin{array}{l} CH_2OH \\ | \\ CHOH \\ | \\ CHOH \\ | \\ CHOH \\ | \\ CO + H_2N.NHC_6H_5 \\ | \\ CH \cdot N.NHC_6H_5 \end{array} \quad = \quad \begin{array}{l} CH_2OH \\ | \\ CHOH \\ | \\ CHOH \\ | \\ CHOH \\ | \\ C:N.NHC_6H_5 \\ | \\ CH:N.NHC_6H_5 \\ \text{Glucosazone.} \end{array} + H_2O$$

It was largely owing to the possibility of forming crystalline osazones from difficultly crystallisable sugars that Emil Fischer was enabled to complete his work on the chemistry of this group of compounds.

(*d*) *Reactions with Silver Nitrate and Cobalt Nitrate.*— In addition to reducing alkaline copper tartrate to cuprous oxide, other alkaline solutions of metallic salts (*e.g.*, Ag, Bi), undergo an analogous change on treatment with certain sugars; for example, ammoniacal silver nitrate is reduced to the metallic state (silver mirror) by glucose and fructose, although cane sugar produces no such reduction. Again, on adding 15 c.c. of a sugar solution to 5 c.c. of a 5 per cent. solution of cobalt nitrate and mixing with 2 c.c. of a 50 per cent.

solution of caustic soda, distinctive colorations or precipitates are produced. Cane sugar gives a violet solution which retains its colour even on boiling, glucose and fructose produce a blue coloration and a precipitate.

The sugars also differ from each other with regard to their action on heating with caustic-soda solution, the readiness with which they are fermentable by yeast, and especially by their different effects upon polarised light when a beam of the latter is passed through an aqueous solution of the sugar (see **17**).

## **17.** *Quantitative Estimation of Sugars by Polarimeter.*

All the sugars give, on dissolving in water, solutions that are optically active ; that is to say, a beam of plane polarised light on being passed through such a solution has its plane of polarisation rotated either to the right or to the left.

The extent of such rotation depends upon several factors :—(i.) The nature of the sugar ; (ii.) the strength of the sugar solution ; (iii.) the length of the column of solution through which the light is passed ; (iv.) the temperature of the solution ; (v.) the wave length (colour) of the polarised light employed.

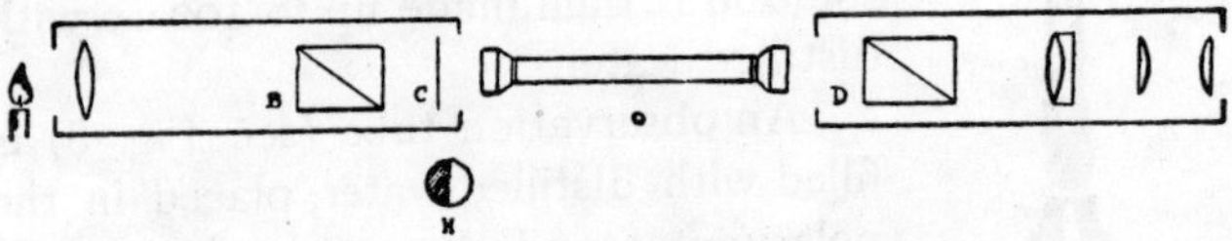


FIG. 7.—Diagramm ic representation of a polarimeter.

These factors are taken into account in calculating the *specific rotation* of a substance, thus :—

The *specific rotation* being defined as the angle of rotation produced by a liquid which in the volume of 1 c.c. contains 1 gram of the active substance, the length of the column through which the light passes being 1 decimetre, we have

$$[a] = \frac{100 . a}{l . c}$$

where $c$ is the number of grams of active substance in 100 c.c. of the solution, $l$ is the length of the column of liquid in decimetres, and $a$ is the actual rotation produced. It is usual to employ sodium light (D line) in polarimetric observations, and the *specific rotation* with regard to light of this wave-length is generally expressed thus :—$[a]_D$. Dextro-rotation is expressed by the positive sign, and lævo-rotation by the negative.

A diagrammatic representation of a polarimeter (Fig. 7) is appended, but for a fuller description of the various types, the different half-shadow arrangements, and the methods of graduating and adjusting the instruments, a work on polarimetric measurements should be consulted. It is absolutely essential, however, for acquiring a thorough knowledge of the method, that the instrument be examined and used practically.

FIG. 8.—Polarimeter observation tube.

**18.** *Determination of Specific Rotation of Cane Sugar.*

Twenty grams of pure sucrose are accurately weighed out, placed in a 100 c.c. graduated flask, and dissolved in distilled water. The volume of the solution is then made up to 100 c.c. with distilled water.

An observation tube (see Fig. 8) is filled with distilled water, placed in the polarimeter, and the zero point on the scale thus determined. The observation tube is then filled with the sugar solution, and the angle of rotation determined several different times, approaching the position of equal illumination of the half-shadow from either side alternately.

From the angle of rotation $a$, the specific rotation is calculated from

$$[a] = \frac{100.a}{l.c} \text{ (see p. 31).}$$

*Specific Rotations of different Sugars.*

| | |
|---|---|
| Cane sugar (sucrose) | $+66{\cdot}5^\circ$ |
| Glucose | $+53^\circ$[1] |
| Fructose | $-71^\circ$ |
| Maltose | $+141^\circ$ |
| Lactose | $+52^\circ$ |

The polarimeter method is employed very largely for the estimation of (cane) sugar in solutions obtained during the manufacture of that substance from cane or beet. The method is capable of great accuracy, and has the additional advantage of being quickly performed.

## 19. *Inversion of Cane Sugar.*

The bioses, being produced from two monose molecules by condensation with the loss of 1 molecule of water, are generally readily convertible into monoses by hydrolysis with dilute acids.

Cane sugar, for example, readily undergoes this change, giving a mixture of equal molecules of glucose and fructose, known as "invert" sugar.

$$\underset{\text{Cane sugar}}{C_{12}H_{22}O_{11}} + H_2O = \underset{\text{Glucose}}{C_6H_{12}O_6} + \underset{\text{Fructose.}}{C_6H_{12}O_6}.$$

The specific rotation of cane sugar is $+66{\cdot}5^\circ$, while that of glucose is $+53^\circ$, and that of fructose $-71^\circ$. The specific rotation of the solution of invert sugar is therefore negative, and consequently the progress of the inversion can be followed directly by means of the polarimeter.

Twenty-five c.c. of the 20 per cent. sugar solution are thoroughly mixed with an equal volume of normal HCl or $H_2SO_4$, and the observation tube of the polarimeter filled as quickly as possible with the mixture.

A reading of the rotation is taken at once and the time of reading noted. As the angle alters quickly during the first few minutes, readings should at first

[1] When freshly dissolved in cold water, glucose gives a much greater rotation, owing to the existence of an unstable isomeride, which passes into a stable equilibrium mixture, slowly in the cold, quickly on heating, or in the presence of alkali.

be taken every two minutes, and then at longer intervals of time as the rate of change decreases.

A curve should then be plotted showing the relation between time and angle of rotation, the last time of reading being about twenty-four hours after the commencement of the inversion.

The following results of an actual experiment are given :—

| Time. | Reading. |
|---|---|
| 0 hrs. 0 mins. | +13° 42′. |
| 0 „ 15 „ | +13° 24′. |
| 0 „ 30 „ | +12° 54′. |
| 0 „ 50 „ | +12° 6′. |
| 1 „ 10 „ | +11° 30′. |
| 1 „ 32 „ | 10° 18′. |
| 2 „ 0 „ | 10° 0′. |
| 2 „ 15 „ | 9° 24′. |
| 3 „ 34 „ | 7° 30′. |
| 5 „ 33 „ | 5° 6′. |
| 7 „ 7 „ | 4° 3′. |
| 8 „ 27 „ | 3° 18′. |
| 24 „ 56 „ | 0° 36′. |
| 27 „ 43 „ | 0° 18′. |

The inversion of cane sugar being brought about by the action of hydrogen ions, and the rate of inversion depending on the concentration of hydrogen ions, a method similar to the above can be used to determine the percentages of ionisation (strengths) of different acids at various concentrations.

## *Starch* $(C_6H_{10}O_5)_n$.

Starch is produced in the leaves of all green plants, and is stored in the seeds, roots, and tubers as a reserve food material for the young plant before it is sufficiently far advanced in growth to perform the functions of assimilation.

Some plant products, such as rice (grain) and potatoes (tubers), contain a sufficient amount of starch to serve as sources of this substance and are largely grown for the purpose.

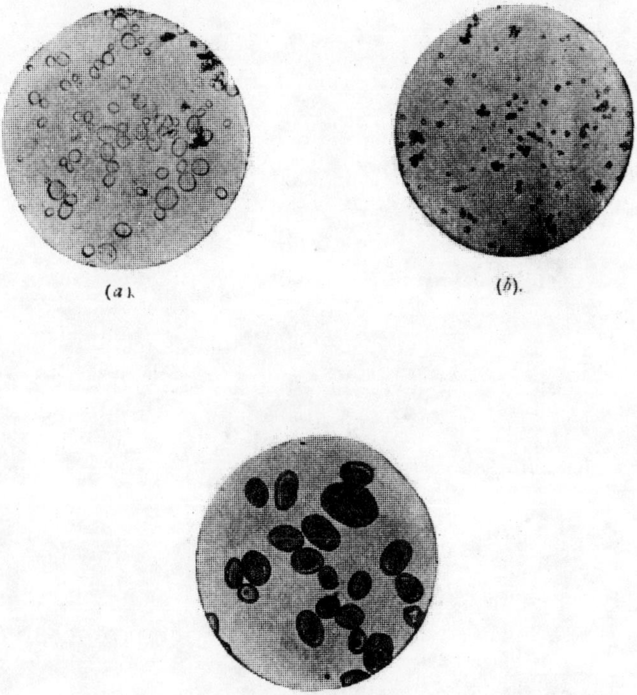

(*a*). (*b*). (*c*).

FIG. 9.—Photomicrographs of starch granules. (Magnification 110.)

(*a*). Wheat starch.
(*b*). Maize starch.
(*c*). Potato starch.

[*Face p.* 84.

Starch exists in plants in the form of granules, which vary in shape and size according to the source. Potato-starch granules are largest in size (·07 to ·02 mm. diameter), while amongst the smallest-sized granules are those of rice starch (·097 to ·005 mm.). Wheat starch is circular in outline, while maize starch is angular, and from a microscopical examination of a sample of starch it is possible to infer the plant from which it was obtained (see Fig. 9).

*Percentages of Pure Starch in various Plant Products (Dry matter).*

| | | | |
|---|---|---|---|
| Potatoes | 80 | Oats | 55 |
| Rice | 78 | Wheat bran | 8 |
| Wheat | 67 | Straw, less than | 1 |

**20.** *Preparation and Qualitative Examination of Starch.*—One or two cleaned potatoes are grated to a pulp, the latter tied in a cloth, dipped in water, and squeezed into a large beaker of water, the cloth and its contents being wetted from time to time. The starch granules pass through the cloth, and render the water in the beaker milky. The starch is allowed to settle to the bottom, and is then washed once or twice by decantation. It is then placed on a porous tile and dried in the steam oven.

The white substance so obtained consists of a number of starch granules, which are made up of a cell wall of starch cellulose (insoluble in water) enclosing a mass of starch granulose or soluble starch (soluble in water).

The following reactions should be performed with starch :—

(i.) A little starch is rubbed into a paste with water, and a few cubic centimetres of the milky liquid poured into a large bulk of water and boiled. The opalescent liquid obtained after boiling consists of a sticky solution of starch granulose containing in suspension the insoluble starch cellulose formed from the cell walls, which have burst under the action of heat and liberated the soluble starch.

(ii.) The boiled solution is filtered, and alcohol added to the clear filtrate. A precipitation of starch granulose takes place.

(iii.) A drop of iodine in potassium-iodide solution is added to the boiled filtered solution, and also to a portion of the original paste. The granulose alone gives the characteristic *blue* coloration of starch iodide. The coloration is discharged on heating to 80° C., but returns on cooling, becoming deeper the lower the temperature.

(iv.) A little starch paste is boiled for a quarter of an hour with dilute HCl, whereby it is hydrolysed, first to dextrin, then to glucose, the presence of which can be demonstrated by addition of Fehling's solution, and boiling.

$$(C_6H_{10}O_5)_n + nH_2O = nC_6H_{12}O_6.$$

(v.) Starch is hydrolysed to maltose in a similar way by the action of diastase (see **34**, p. 55).

# CHAPTER III

## PROXIMATE CONSTITUENTS OF PLANTS—*continued*

### (*e*) **The Alkaloids.**

THE alkaloids are nitrogenous organic substances occurring in small proportions in certain plants, generally of the dicotyledonous family. Their importance lies in the fact that they are bodies of great therapeutical value, owing to their marked action on certain nerve centres. They are generally powerful poisons, when taken in any except minute doses.

Although the majority of the alkaloids contain oxygen in their composition, some, *e.g.*, nicotine and conine, are free from this element, and these are usually liquid, while the others are solid crystalline substances.

In the plants in which they occur, the alkaloids are generally found in largest amount in the seeds or the leaves.

Being secondary or tertiary amines, they are all basic in character, and combine with acids to form salts, and it is often in the state of salts that the alkaloids exist in plants; they are generally optically active, although a few—piperine, for example—have no action upon the plane of polarisation of light.

A few of the more important alkaloids, and the plants in which they occur, and their distribution in the plant, are given herewith:—

| Alkaloid. | Plant. | Greatest Proportion of Alkaloid in |
|---|---|---|
| Quinine | Cinchona | Bark |
| Nicotine | Tobacco | Leaves |
| Strychnine | Nux vomica bean | Seeds |
| Conine | Hemlock | Seeds |
| Caffeine | Tea, coffee | Leaves |
| Piperine | Pepper | Seeds |
| Atropine | Deadly nightshade | Seeds, leaves |
| Cocaine | Coca plant | Leaves |
| Morphine | Poppy (special varieties) | Seed capsules |

The extraction of the alkaloids and their methods of estimation differ greatly, but in order to give some idea of the processes, the extraction of caffeine from tea and the estimation of nicotine in tobacco leaves are given below.

### 21. *Extraction of Caffeine from Tea.*

A hundred and fifty grams of tea are boiled with about 700 c.c. water for a quarter of an hour, and the infusion is then filtered through unsized calico into a large porcelain basin or beaker. The leaves remaining on the cloth are then thoroughly washed with 300 c.c. boiling water, the washings being allowed to run into the liquid already obtained. The resulting liquid consists of an aqueous solution of caffeine, tannin, albuminous substances, and colouring matter. To remove the tannin and albuminous substances, the solution in the beaker is kept hot over a Bunsen burner, and basic lead-acetate solution[1] added until no further precipitation takes place. The liquid is at once filtered as quickly as possible, and the precipitate washed with a small quantity of hot water. The excess of lead is then removed by addition of sufficient dilute $H_2SO_4$, the precipitated lead sulphate being removed by decantation.

To remove the colouring matter, 4 or 5 grams of animal charcoal are stirred into the liquid, which is then concentrated down to about 300 c.c. The solution is filtered, and the treatment with animal charcoal repeated if necessary.

The caffeine is extracted from the filtrate by shaking in a separating funnel with 70 c.c. chloroform. This extraction should be repeated at least three times.

The chloroform is then distilled off on the water-bath, and the residue of caffeine taken up with a small quantity of very hot water. The caffeine slowly separates from this aqueous solution in the form of white silky needles.

Caffeine, $C_8H_{10}N_4O_2$, is a silky crystalline substance melting at 233° C. It differs from the other alkaloids in being a derivative of purine, and hence related to uric acid.

**22.** *Estimation of Nicotine in Tobacco.*

**The method devised by Kissling depends upon the fact that nicotine in the form of the free base is volatile in steam, and can be distilled over to give an aqueous solution which is alkaline and can be directly titrated with acids.**

**[1] Basic lead acetate can be previously prepared by boiling a solution of lead acetate with excess of PbO, and then filtering.**

Twenty grams or so of tobacco or tobacco leaves are dried at a temperature of 60° and are finely ground while still warm, preferably in a hand-mill. Ten grams of the powder are placed in a small beaker and 10 c.c. of alcoholic-soda solution added.[1] The object of this addition of caustic soda is to liberate the free nicotine from any salt, such as malate, citrate, etc., in which it may exist combined in the leaves. The mixture is well stirred with a spatula, and when homogeneous is placed in a filter-paper cartridge and extracted with ether in a Soxhlet extractor for five hours.

The ether is then distilled from the extract on the water-bath, and the residue containing all the nicotine taken up with 50 c.c. of a ·4 per cent. aqueous solution of caustic soda. The green liquid is transferred to a 500 c.c. flask and distilled in steam, with very efficient condensation (Fig. 10). The distillation is carried on till about 400 c.c. have collected in the receiver, when all the nicotine should have passed over. The distillate is titrated with $\frac{N}{2}$ $H_2SO_4$, using cochineal as indicator, and from the amount of the standard acid used the percentage of nicotine in the leaves can be readily calculated.

One c.c. $\frac{N}{2}$ $H_2SO_4$ is equivalent to ·081 gram nicotine.

Nicotine, $C_{10}H_{14}N_2$, is a colourless liquid, and is one of the few alkaloids containing no oxygen. It is strongly lœvorotatory $[\alpha]_D = -162°$, boils at 247° C., and is slowly oxidised by air to a dark liquid

---

[1] The alcoholic-soda solution is prepared by dissolving 6 grams caustic soda in 40 c.c. distilled water and 60 c.c. 90 per cent. alcohol.

Ten c.c. of this solution contains caustic soda more than sufficient in amount to liberate all the combined nicotine present in 10 grams of leaves containing up to 10 per cent. nicotine.

with an unpleasant odour, recalling that of a foul tobacco pipe. It is a powerful poison, and is used largely in horticulture as a most efficient spray against certain insect pests, such as the different forms of aphis affecting apples, damsons, hops, and roses.

A solution containing ·075 per cent. of nicotine is employed for spraying purposes.

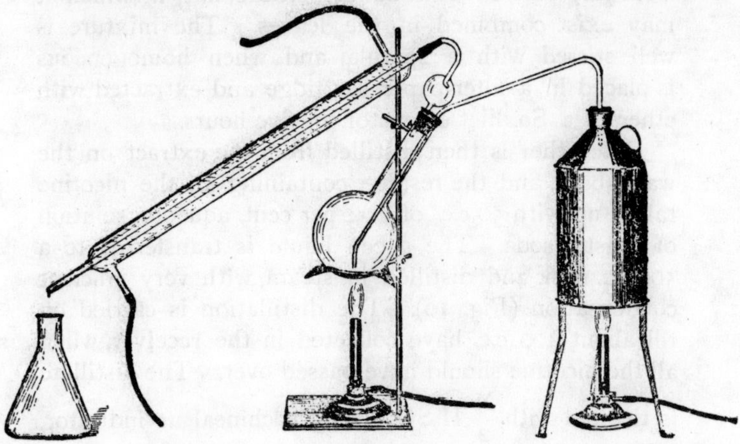

FIG. 10.—Apparatus for distillation with steam.

## 23. *Qualitative Reactions of the Alkaloids.*

The alkaloids, being of aminoid constitution, are characterised by forming insoluble precipitates with certain reagents mostly of an acid nature. These general precipitants may be conveniently used for the separation of members of the alkaloids from solutions containing them, the alkaloids being afterwards freed from the precipitates by the action of caustic or carbonated alkalies.

The general precipitants for the alkaloids comprise the following :—

Tannic acid, giving a white or yellowish precipitate.

Picric acid, giving a yellow crystalline precipitate.

Phosphomolybdic acid, giving a buff or yellowish precipitate.

Potassium mercuriodide (Mayer's reagent), giving a white precipitate.

Potassium tri-iodide, giving a brown precipitate.

Gold chloride, giving a yellow precipitate.

In addition to the above general precipitants, the alkaloids may be distinguished from each other by means of certain distinctive colour tests.

The general colour reagents most usually employed in the case of the commoner alkaloids are the following :—

Concentrated $HNO_3$; concentrated $H_2SO_4$; and concentrated $H_2SO_4$ together with a crystal of potassium bichromate; while Erdmann's reagent (concentrated $H_2SO_4$, containing ·01 per cent. concentrated $HNO_3$), Froehde's reagent (1 per cent. solution of $(NH_4)_2 MoO_4$ in concentrated $H_2SO_4$), and Mandelin's reagent (·5 per cent. solution of vanadium chloride in concentrated $H_2SO_4$) afford characteristic colour reactions in certain cases.

The reactions given with certain of the commoner alkaloids are tabulated herewith :—

| | Concentrated $HNO_3$. | Concentrated $H_2SO_4$. | $H_2SO_4$ and $K_2Cr_2O_7$. |
|---|---|---|---|
| Morphine . | Orange red colour . | Pinkish becoming brown | Greenish brown |
| Quinine . | Light yellow colour | Colourless . . . | Bright green |
| Strychnine . | Yellow colour . | Colourless . . . | Violet |
| Caffeine . | Light yellow colour | Colourless . . . | Green (gradual) |
| Nicotine . | Brown colour . . | Light brown . . . | Light brown |

Special confirming tests are as follows :—

*Morphine.*—(i.) A small quantity of formaldehyde added to the alkaloid or one of its salts, followed by two or three drops of strong $H_2SO_4$ produces an intense purple colour.

(ii.) Froedhe's reagent (see above) gives a violet colour changing to dirty green.

*Quinine.*—(i.) Excess of dilute $H_2SO_4$ gives a marked blue fluorescence.

(ii.) Chlorine water or bromine water followed by ammonia gives a green colour.

(iii.) On moistening in an evaporating dish with strong HCl, and warming over a Bunsen flame, a violet colour and violet vapours are observed just before charring takes place.

*Strychnine.*—Mandelin's reagent (see above) gives a bright violet coloration.

*Caffeine.*—On evaporating to dryness with a little bromine water, and moistening the yellow residue with $NH_4OH$, a bright purple coloration is produced (Murexide test).

*Nicotine.*—(i.) Formaldehyde and then nitric acid gives a bright rose colour.

(ii.) $HgCl_2$ solution gives a white precipitate turning yellow on standing.

### (*f*) The Glucosides.

The glucosides are crystalline substances occurring in plants, and which on hydrolysis, either by accompanying enzymes, or by dilute acids, give various organic compounds together with glucose (hence the name glucoside). Some glucosides give also hydrocyanic acid, and in this case the hydrolysis is called cyanogenesis and the glucosides cyanogenetic. For example, the glucoside of bitter almonds, *amygdalin*, is hydrolysed by the accompanying enzyme *emulsin*, to give glucose, hydrocyanic acid, and benzaldehyde.

A cyanogenetic glucoside is also found in linseed and linseed cake (see section on Feeding-Stuffs, p. 155).

As examples of non-cyanogenetic glucosides, salicin (occurring in willow bark) and digitalin (occurring in seeds of the purple foxglove) may be mentioned.

**24.** *Hydrolysis of Salicin.*

About 2 grams of salicin are boiled with dilute sulphuric acid for a few minutes, whereby hydrolysis takes place to glucose and saligenin (*o*-hydroxybenzyl alcohol):

$$C_{13}H_{18}O_7 + H_2O = C_6H_4\begin{cases} CH_2OH \\ OH \end{cases} + C_6H_{12}O_6.$$

The following tests should be performed with the solution to demonstrate the presence of the products of hydrolysis :—

(i.) The hydrolysed solution is rendered alkaline with caustic soda, Fehling's solution added, and the mixture boiled. The characteristic red precipitate of cuprous oxide shows the presence of glucose.

(ii.) A small quantity of strong $K_2Cr_2O_7$ solution is added to the hydrolysed solution, which is then boiled with 1 or 2 c.c. of concentrated sulphuric acid. The saligenin is oxidised to salicylic aldehyde, which is recognised by its characteristic smell, somewhat resembling that of benzaldehyde.

**25.** *Hydrolysis of Amygdalin.* (Cyanogenesis.)

About 20 grams of almonds are ground up, soaked in water, and allowed to stand for half an hour.

The enzyme emulsin hydrolyses the amygdalin in the presence of water, with production of glucose, benzaldehyde (oil of bitter almonds), and hydrocyanic acid, thus :—

$$C_{20}H_{27}NO_{11} + 2H_2O = C_6H_5CHO + HCN + 2C_6H_{12}O_6.$$

The benzaldehyde is recognised by reason of its smell, while the presence of hydrocyanic acid can be demonstrated as follows :—

The liquid is transferred to a distillation flask, and about 20 c.c. distilled over into a receiver through a suitable condenser.

The following tests for hydrocyanic acid are then applied to the distillate :—

(i.) A silver nitrate solution produces a white precipitate of AgCN.

(ii.) A little of the solution is placed in a basin together with a few drops of yellow ammonium sulphide, and the mixture evaporated nearly to dryness. A thiocyanate (sulphocyanide) is produced as shown by the addition of a drop of ferric chloride solution, when the characteristic blood-red coloration of ferric thiocyanate is formed.

(iii.) A little caustic soda solution is added, then a drop or two each of ferric chloride and ferrous sulphate solutions. On warming and acidifying with hydrochloric acid, a dark blue precipitate of Prussian blue (ferric ferrocyanide) is produced.

By the same method as that described above, the presence of amygdalin can be shown in cherry-laurel leaves, and in the kernels of peach, plum, and cherry stones.

### (*g*) The Organic Acids.

Various organic acids—formic, oxalic, malic, citric, tartaric, etc. —occur very widely distributed in the vegetable kingdom, and are hence sometimes known as the vegetable acids.

They are generally present in the form of metallic salts, especially those of potassium or calcium, but they sometimes occur in the free state, as for example in unripe fruits, which owe their acidity to this cause, and also to the presence of acid salts of potassium and calcium.

Some of the commoner organic acids and their salts occurring in plants are enumerated :—

*Formic acid*, H . COOH, in nettles.

*Acid potassium oxalate*, COOH . COOK, in sorrel, rhubarb.

*Malic acid*, COOH . CHOH . $CH_2$ . COOH, in apples, mountain-ash berries, gooseberries, morella cherries.

*Acid potassium malate*, COOK . CHOH . $CH_2$ . COOH, in sweet cherries.

*Acid potassium tartrate*, COOK.CHOH.CHOH.COOH, in grapes.

*Citric acid*, COOH.$CH_2C(OH)(COOH).CH_2$.COOH, in lemons, sour gooseberries.

The organic acids present in unripe fruit are converted into other non-acid bodies during the process of ripening.

## 26. *Preparation and Qualitative Examination of Crystallised Oxalic Acid* $(COOH)_2.2H_2O$.

Oxalic acid can be conveniently prepared in the laboratory by the oxidation of cane sugar with nitric acid.

About twenty-five grams of cane sugar are dissolved in 130 c.c. water and 100 c.c. concentrated $HNO_3$ added. The mixture is gently warmed and the heating discontinued when oxidation commences, as evidenced by the rapid production of brown fumes of oxides of nitrogen.

When the reaction slackens, the liquid is placed on the water-bath, or gently boiled, until its volume is reduced by about one-half. It is then placed aside to cool and crystallise, and may with advantage be seeded with a ready-formed crystal of oxalic acid.

The oxalic acid is recrystallised from small quantities of water until free from nitric acid.

The following tests for oxalic acid should be performed with the crystals:—

(i.) On heating the crystallised acid in a dry tube, it melts in its own water of crystallisation, and on further heating sublimes.

(ii.) On heating oxalic acid or an oxalate with concentrated $H_2SO_4$, CO and $CO_2$ are produced by dehydration of the oxalic acid—

$$\begin{array}{l} COOH \\ | \\ COOH \end{array} = CO_2 + CO + H_2O$$

The presence of $CO_2$ can be demonstrated by its action on lime water, while the CO will burn at the mouth of the test-tube with its characteristic blue flame.

(iii.) A neutral solution of ammonium oxalate is prepared by adding a slight excess of $NH_3$ to an aqueous solution of the acid, and boiling for a few seconds.

On adding a solution of $CaCl_2$ to this neutral solution, a white crystalline precipitate of calcium oxalate is produced, insoluble in acetic acid, but soluble in HCl or $HNO_3$.

(iv.) On adding a solution of an oxalate or oxalic acid to an acidified solution of potassium permanganate, the colour of the permanganate is discharged, owing to reduction by the oxalic acid—

$$2KMnO_4 + 5H_2C_2O_4 + 3H_2SO_4 = K_2SO_4 + 2MnSO_4 + 10CO_2 + 8H_2O.$$

*The presence of oxalic acid in plants* may be conveniently shown in the case of sorrel (*oxalis acetosella*), in which it exists in the form of potassium hydrogen oxalate and as calcium oxalate.

About 25 grams of sorrel are extracted by grinding up with warm water, and the solution filtered off. The liquid is boiled down to a small volume, and tests applied as in (iii.) and (iv.) above.

## 27. *Qualitative Reactions of Tartaric Acid and Tartrates.*

Tartaric acid, COOH . CHOH . CHOH . COOH, exists in the form of large colourless transparent crystals readily soluble in water. It is a dibasic acid, and therefore forms acid and normal salts. It exists in four stereo-isomeric forms, two of which are optically active (dextro and lœvo) and two optically inactive.

(i.) On heating in a dry tube, tartaric acid and tartrates char readily, with production of inflammable vapours and a smell of burnt sugar. A residue of oxide or carbonate is left in the case of metallic tartrates.

(ii.) On heating in a test-tube with concentrated $H_2SO_4$, immediate charring takes place. The tartrate is oxidised with production of CO and $CO_2$, whilst the sulphuric acid undergoes reduction to $SO_2$, all of which three gases may be recognised by the usual tests.

(iii.) Metallic tartrates reduce ammoniacal silver oxide to the metallic state, and under suitable conditions the metal is deposited in the form of a silver mirror.

To obtain the best results this test should be carried out as follows. To a neutral solution of the tartrate, silver nitrate is added until a decided white crystalline precipitate of silver tartrate is produced. Very dilute ammonia solution is then added until the precipitate is nearly dissolved, and the mixture is then heated by immersing the test-tube in a beaker of boiling water. Care should be taken not to move the test-tube during the heating.

(iv.) Calcium chloride solution added in small quantities to a solution of tartaric acid gives no precipitate, owing to the formation of soluble acid calcium tartrate. On adding a further quantity of the calcium salt, however, the insoluble normal tartrate is precipitated. This precipitate is soluble in tartaric acid and also in acetic acid.

(v.) A small crystal of solid tartaric acid, or a tartrate, and an equal quantity of pyrogallic acid are warmed with 1 or 2 c.c. of concentrated $H_2SO_4$ in a test-tube. A fine violet-blue coloration is produced (distinction from citrate).

## 28. *Preparation of Citric Acid from Lemons.*

Citric acid occurs in lemons to an extent of about 8 per cent., and can be readily prepared from the juice.

The citric acid is neutralised by chalk, and the calcium citrate formed is precipitated by boiling. The calculated quantity of dilute $H_2SO_4$ is then added to the calcium citrate, the calcium sulphate so formed is removed, and the solution of citric acid is evaporated until it crystallises.

Three or four lemons are cut in half and the juice squeezed into a small beaker. The liquid is diluted with about an equal volume of water, and the solution boiled for some time to coagulate albuminous substances. It is then filtered while hot.

The volume of the liquid is carefully measured by means of a graduated flask (100 c.c.) and a burette, an aliquot part (10 c.c.) removed, and the citric acid present in it estimated by titration with $\frac{N}{2}$ NaOH solution, using phenolphthalein as indicator.

The amount of standard $H_2SO_4$ equivalent to the citric acid in the remaining volume of juice is then calculated. This quantity of $H_2SO_4$ is that added to the calcium citrate in a later stage of the preparation.

Powdered chalk is then stirred into the liquid in small quantities at a time, excessive frothing being avoided, until a slight excess of the chalk has been added, as rendered evident by the fact that some of the chalk remains undissolved at the bottom of the vessel.

The solution containing calcium citrate is poured off from the undissolved chalk, and boiled vigorously for a few minutes. The calcium citrate is thus rendered insoluble, and falls as a thick white crystalline precipitate.

This precipitate is filtered off at the pump on a Büchner funnel, and well washed with boiling water.

The washed calcium citrate is transferred, while still wet, to a beaker, and the calculated quantity of standard acid (above) run on to it from a burette.

The solution is then boiled and filtered from calcium sulphate. To remove the sulphate remaining in solution, the solution is concentrated by boiling until calcium sulphate commences to separate out, and then an equal volume of absolute alcohol is added to complete the precipitation. The liquid is again filtered, the clear solution obtained being one of citric acid in dilute alcohol. The liquid is evaporated on the water-bath until it attains the consistency of a thick syrup, and is then set aside to crystallise.

As this syrupy solution of citric acid shows the phenomenon of supersaturation to a marked degree, it is advisable to hasten the crystallisation by seeding with a small crystal of citric acid. The citric acid crystals obtained after some time are freed from mother liquor by pressing between filter papers.

## 29. *Qualitative Reactions of Citric Acid.*

Citric acid forms colourless crystals containing one molecule of water of crystallisation. It readily dissolves in water and alcohol, and forms three classes of metallic salts, being a tribasic acid.

The following tests should be applied to the citric acid obtained in the above preparation :—

(i.) On heating in a dry tube, citric acid or citrates char, with evolution of acid vapours.

(ii.) On heating with concentrated $H_2SO_4$, CO and $CO_2$ are evolved, but charring does not take place immediately (distinction from tartrate).

(iii.) Calcium chloride, if moderately dilute, gives no precipitate with a neutral solution of a citrate in the cold. On heating, however, calcium citrate is precipitated as a white crystalline precipitate.

(iv.) Silver nitrate produces with a neutral solution

of a citrate, a white precipitate of silver citrate, soluble (like the tartrate) in ammonia. On being warmed, however, the ammoniacal solution does not deposit metallic silver (distinction from tartrate).

### (*h*) The Essential Oils.

Classed together under the generic term of essential oils, are a number of plant products that resemble each other in that they are all bodies possessed of a penetrating, generally pleasant aromatic smell, but which differ very largely in chemical constitution. Strictly speaking, the essential oils are not pure chemical substances, but are more often mixtures of different oils. Some are true hydrocarbons (terpenes), while many contain oxygen in their composition, and are ethereal salts, aldehydes, or phenols. The essential oils are all readily volatile, and can be distilled with steam at temperatures much below their boiling-points. They are generally present in plants in only small quantities (less than 1 per cent.) but are of great importance in certain animal food-stuffs (hay) in imparting palatability.

Some of the more important essential oils are the following :—

*Terpenes* (all of formula $C_{10}H_{16}$).

Pinene, in oil of turpentine, oil of eucalyptus.

Limonene, in oil of orange, oil of lemon.

Cymene, in oil of caraway, oil of eucalyptus.

*Ethereal Salts.*

Linolyl acetate, $C_{10}H_{17} . OOC . CH_3$, in oil of lavender.

Allyl isothiocyanate, $C_3H_5 . CNS$, in oil of mustard.

Allyl sulphide, $(C_3H_5)_2S$, in oil of garlic.

Propyl acetate, $CH_3 . COOC_3H_7$, in essence of pears.

Ethyl butyrate, $C_3H_7 . COOC_2H_5$, in essence of pineapple.

*Aldehydes.*

Cinnamic aldehyde, $C_6H_5 . CH : CH . CHO$, in oil of cinnamon.

Benzaldehyde, $C_6H_5 . CHO$, in oil of bitter almonds (see p. 45).

*Phenols.*

Thymol, $C_{10}H_{13}OH$, in oil of thyme.

Carvacrol, $C_{10}H_{13}OH$, in oil of hops and oil of cloves.

The essential oils are generally prepared by distillation of the plants with steam. In the case of the more delicate perfumes (rose, violet, orange-flower), the oil is extracted by macerating with lard.

**30.** *Preparation of Essential Oils by Distillation with Steam.*

Some heads of lavender, pieces of orange-peel, or peppermint leaves are placed in a distilling flask, half-filled with water, and fitted with a condenser and receiver. About one-third of the water is distilled off, whereby a large proportion of the essential oil, being volatile in steam, will be carried over into the distillate.

The distillate is extracted with small quantities of ether in a separating funnel or burette, the ethereal solution separated from the aqueous portion, and the ether distilled off on the water-bath. The residue of essential oil will be found to possess the characteristic odour of the substance taken.

## (*i*) Miscellaneous Substances—Tannins, Resins, Chlorophyll.

*Tannins.*—The tannins are mostly derivatives of gallic acid or trihydroxybenzoic acid, $C_6H_2(OH)_3.COOH$, and occur in plants mainly as glucosides, although the best known and most important member, ordinary tannin or gallotannic acid, exists in the free state. It occurs in comparatively large quantities in oak galls, of which it makes up half the weight, in oak-bark, and in smaller proportions in other plant products, such as tea.

The function of the tannins in the economy of plants is at present unknown, although they are thought to be elaborated from the carbon dioxide of the air, in a somewhat similar way to that in which the sugars and starches are produced. Tannin is of importance owing to its extensive use for the conversion of hides into leather, an action depending on the fact that certain substances (gelatine, etc.) present in the hides are converted into complex insoluble substances impervious to water.

The presence of tannin in food-stuffs is undesirable, as the astringency of the substance leads to a decreased flow of the gastric juices, with the result that digestion is to a certain extent impaired.

## 31. *Qualitative Examination of Tannin* ($C_{14}H_{10}O_9 + 2H_2O$).

Tannin is generally obtained as a light brown semi-crystalline solid, which is white when pure. It dissolves readily in water, giving a solution with an intensely astringent taste; it also dissolves readily in aqueous alcohol, but is insoluble in absolute alcohol or ether.

The following tests should be applied with tannin, and *also* with the solution obtained by extracting a few powdered oak-galls with water, filtering, and concentrating the solution.

(i.) An alkaline solution of tannin (alkaline gallotannate) when exposed to the air turns brown, owing to oxidation.

(ii.) On adding a solution of ferric chloride to a solution of tannin, a black or greenish black coloration or precipitate is produced. This colouring matter forms the basis of ordinary writing (not printer's) ink.

(iii.) Tannin forms a buff-coloured flocculent precipitate with a cold solution of gelatine. A similar action takes place in hides during the process of tanning.

(iv.) Lead acetate gives a white precipitate of lead tannate (gallotannate). The precipitate is decomposed by $H_2S$ to give lead sulphide and tannin, and this reaction can hence be used for the preparation of the latter substance in a pure condition.

### Resins.

The resins are amorphous substances occurring in plants, and are frequently associated with essential oils, from which they appear to be formed by oxidation processes—common "rosin," for example, being produced in such a way from the terpenes present in pine trees.

Many resins, such as guaiacum, gamboge, asafœtida, opoponax, are employed for various purposes, while in the case of the hop the value of the plant depends very largely upon the amount of resins, which occur in this case also in association with a terpene ($C_{10}H_{16}$), "hop-oil." This oil gives the characteristic aroma and flavour to the beer, while the bittering and preservative properties of hops are known to reside in the resins.

### 32. *Quantitative determination of Resins in Hops.*

There are two varieties of resins present in hops: "soft" resins, soluble in petroleum ether, and "hard" resins, which can be extracted with ordinary ether after the removal of the "soft" resins. According to Briant and Meacham, it is the "soft" resins alone that are preservative in their action.

The quantitative estimation of the "soft" resins is performed as follows:—From 3 to 5 grams of hops are weighed out and carefully placed in a cartridge of filter paper that has been previously extracted with petroleum ether. The cartridge and contents are then extracted for twenty-four hours in a Soxhlet apparatus, using petroleum ether of B.P. 40°-50° C.

The ethereal extract is then drained from the cartridge into the extraction flask, and the ether evaporated off in the water-bath, the ether being conveniently condensed and collected in the Soxhlet apparatus, from which the extracted cartridge has been removed. The flask and residue is next transferred to a steam oven, dried for six hours in the steam oven, and weighed.

Approximate percentages of "soft" resins in different varieties of hops are as follows:—

| | | | |
|---|---|---|---|
| Fuggles, . . . | 10 to 11 | Holiedaus, . . . | 16 to 18 |
| East Kent Goldings, . | 11 to 13 | Oregons, . . . | 16 to 17 |
| Canterbury Whitebines, | 12 to 15 | | |

## Chlorophyll.

Chlorophyll is the name given to the green colouring-matter of plants, in the economy of which it plays a highly important *rôle*. The function of chlorophyll is the assimilation of carbon dioxide from the air and its elaboration into cane sugar, and ultimately other sugars, starches, and cellulose. This assimilative action of chlorophyll takes place in conjunction with the protoplasm also present, and proceeds only in the presence of light. The rapidity of the action varies with the colour of the light, being most vigorous under the influence of those rays (red, orange, yellow, blue) which are absorbed by the chlorophyll. (Engelmann.)

Although magnesium is stated to be an invariable constituent of the substance, the chemical structure of the chlorophyll molecule is unknown ; in fact, the substance is probably a mixture of different compounds. The presence of iron in the plant is indispensable for chlorophyll formation, although this element does not enter into its constitution.

Chlorophyll is insoluble in water, but dissolves readily in ether, carbon bisulphide, and alcohol.

### 33. *Preparation and Qualitative Examination of Chlorophyll.*

Twenty grams or so of fresh green leaves are cut into small pieces and bruised in a mortar. They are then placed in a filter-paper cartridge and extracted with ether for an hour in a Soxhlet extractor on the water-bath. The bright green ethereal solution of chlorophyll is evaporated to about one-third of its volume, and the following properties of the resulting solution noted :—

(i.) Chlorophyll solutions show a marked red fluorescence, especially when viewed by reflected light.

(ii.) Chlorophyll is soluble in alcohol, ether, $CS_2$, but insoluble in water. A small quantity of the ethereal solution is shaken up with water in a test-tube, and allowed to stand. It will be seen that all the chlorophyll remains in the top ethereal layer, the lower aqueous layer being only slightly coloured.

(iii.) Chlorophyll solutions on exposure to light rapidly lose their colour, only a faint yellow tinge remaining. This change is due to oxidation.

(iv.) A few drops of $HNO_3$ shaken up with the chlorophyll solution will bring about a similar change.

# CHAPTER IV

## Chemical Changes during Germination and Growth

During the germination of a seed, for which certain conditions of temperature, moisture, and ventilation are necessary, most of the substances in the seed undergo chemical changes, the general trend of such changes being the production of soluble bodies from insoluble substances. These changes are brought about by the action of unorganised ferments or enzymes (cytase, diastase) present in all seeds, and which undergo a considerable increase both in quantity and in activity during the process of germination. The substances rendered soluble are reserve food-materials for the young plant, and they undergo this change in order that they may dissolve in the sap and be transported to those parts of the plant where they are required.

Cellulose is rendered soluble by the cytase, starch is converted into dextrins, and then to maltose by the diastase, insoluble proteins are rendered soluble, and fat is converted into starch and then into maltose.

### 34. *Conversion of Starch to Sugar.*

Forty grains of sound barley are germinated on damp filter paper until the shoot is about an inch long.

They are then ground up in a mortar with a little water, and the solution is filtered. This solution contains maltose, the presence of which can be shown by Fehling's test. The red precipitate of cuprous oxide is produced less easily than in the case of glucose; hence on boiling another portion of the solution for a short time with a few drops of concentrated $H_2SO_4$, and applying Fehling's test, the reduction will be seen to take place with greater rapidity, owing to the fact that the maltose has been hydrolysed to glucose.

Some ungerminated barley should be ground up with water and the solution treated in the same way as above. It will be found that with Fehling's solution a very slight reduction, if any, takes place.

### Photo-synthesis.

The photo-synthesis of the carbohydrates and other organic matter from the carbon dioxide of the atmosphere is probably accomplished by the primary formation of hydrogen peroxide and formaldehyde under the influence of the plant chlorophyll :—

$$CO_2 + 3H_2O = H \cdot CHO + 2H_2O_2.$$

Both of these products are toxic to plants, but the formaldehyde is immediately polymerised to hexose sugars, and the hydrogen peroxide is decomposed by the universally existing plant catalase. Both are thus removed as soon as formed.

### 35. *Plant Catalase.*

The existence of catalase in the green leaf may be shown as follows:—About 6 grams of fresh leaves are pounded in a mortar with a little sand, transferred to a small flask, and mixed with 25 c.c. of 20-volume hydrogen peroxide. The flask is completely filled with water and fitted with a stopper carrying a delivery tube leading under the surface of water, a test-tube being inverted over the end. The delivery tube is also filled with water, and the whole apparatus allowed to stand in a warm place for some time. Oxygen is evolved and is collected in the test-tube, and may be recognised by the ordinary tests.

The experiment should be repeated with a similar quantity of leaves which have previously been plunged for a short time into boiling water, whereby the catalase is destroyed.

# SECTION II.—SOILS

## CHAPTER V

### PROXIMATE CONSTITUENTS OF SOILS

**Limestone, Humus, Sand, and Clay.**

SOILS are popularly classified as loams, clays, marls, peaty, sandy, calcareous, etc., according to their content of the proximate constituents, *sand*, *clay*, *limestone*, *and humus*.

**36.** *Limestone.*

(*a*) The limestone or calcium carbonate content of the soil must be sufficient to ensure a permanently neutral reaction. A rough idea of the amount of $CaCO_3$ present in a soil may be obtained as follows:—

A small quantity of soil (about 5 grams) is dried and powdered, placed in a beaker, and treated with some concentrated hydrochloric acid. If a brisk effervescence is produced the soil contains over 1 per cent. of $CaCO_3$, which, in general, is ample for fertility. Under 1 per cent. is indicated by an evolution of $CO_2$ which can only just be seen. Less than ·5 per cent. can only be detected by holding the containing beaker to the ear and listening for the bursting of the small bubbles of carbon dioxide. If no noise is heard in this way the soil contains practically no limestone, and should be tested for acidity with litmus paper.

(*b*) *Estimation of $CO_2$ in Limestone.*—This estimation may conveniently be carried out in Schrötter's apparatus (Fig. 11), which consists essentially of a flask with three openings. A is closed with a glass stopper and is used for introducing the material. B is in reality a tap-funnel, used for running in the dilute acid, the delivery tube being extended and bent round to prevent splashing. The outlet tube C is half-filled with concentrated sulphuric acid, which thoroughly dries the liberated carbon dioxide before allowing it to escape.

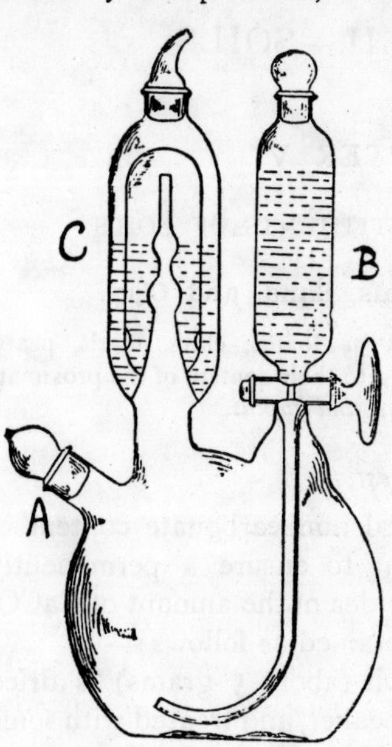


FIG. 11.—Schrötter's apparatus for estimating carbonates.

*Procedure.* — The dropping funnel B is filled with dilute nitric acid and the apparatus weighed. About 1 gram of the carbonate is introduced into the flask, the stoppers are inserted, and the apparatus again weighed. Acid from the funnel is now slowly admitted, so that the liberated $CO_2$ escapes in single bubbles, which pass through the drying tube at a rate of about two per second. When the evolution of the carbon dioxide has ceased, the remainder of the acid is run in, the tap left open, and the tube of the drying apparatus connected to an aspirator of some sort, or simply sucked carefully with the mouth. During aspiration, which must not be too long continued, the apparatus and its contents are carefully heated to incipient boiling. The flask is then allowed to cool, the stoppers replaced, and the whole apparatus weighed. The loss of weight gives the amount of carbon dioxide in the weight of limestone taken.

A simplified form of apparatus is shown in Fig. 12. The dilute acid is kept in the small test-tube which

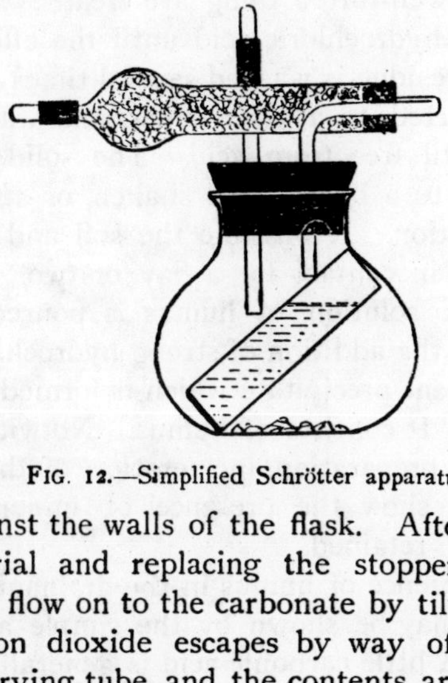

FIG. 12.—Simplified Schrötter apparatus.

leans against the walls of the flask. After introducing the material and replacing the stopper, the acid is caused to flow on to the carbonate by tilting the flask. The carbon dioxide escapes by way of the calcium chloride drying tube, and the contents are then heated and aspirated as before.

This method and apparatus may be used for estimating limestone in very chalky soils, but more accurate methods of estimating small quantities of carbonate present in soil are necessary, and are described under Soil Analysis (**75**, **76**, pp. 94 and 96).

**37.** *Humus.*

(*a*) About 5 grams of air-dried soil are carefully weighed out and heated in the steam oven until of constant weight. The loss in weight is *moisture*. The dry residue is carefully ignited in a platinum or porcelain dish; when cold, it is again weighed. The loss is partly due to the liberation of combined water, but mainly to the decomposition of organic matter, chiefly humus.

(*b*) *Preparation of Humus.* — About 50 grams of peaty soil or well-rotted dung are treated with dilute (10 per cent.) hydrochloric acid until the effervescence ceases. The residue is washed several times by decantation, transferred to a filter paper, and washed with hot water until free from acid. The solid matter is then removed to a beaker and shaken, or stirred, with ammonia solution. If possible the soil and ammonia should be left in contact for a day or two. In either case the black solution of humus is poured off and neutralised by the addition of strong hydrochloric acid.

The flocculent precipitate which is formed is filtered off and dried. It consists of humus. Notwithstanding its method of preparation, incineration of the product will generally show the presence of mineral matter which has been retained.

(*c*) The presence of humus in the drainage from the manure-heap may be shown by the simple addition of dilute HCl. A little carbonic acid is generally evolved, and the humic acid is precipitated as above.

The chemical composition of humus is indefinite, and varies very considerably. It contains, as a rule, substances of the type of humic and ulmic acids, in the form of salts of calcium, and these must be decomposed before the humus can be separated.

The water-retentive power of humus is greater than that of clay or sand, and the value of humus from an agricultural point of view largely depends upon this property of water retention.

**38.** *Sand.*

(*a*) Ten grams of the soil are weighed out, transferred to a beaker, and shaken up with a large bulk of water containing a little caustic soda. The mixture is stirred well, allowed to stand for several minutes, and the supernatant liquid poured off. The sediment which remains behind is twice more treated in the same manner, the several washings being retained for the subsequent separation of the clay (**39** *a*). The liquid should eventually pour off perfectly clear. The residual matter is washed into a weighed dish, dried, ignited, and weighed. It consists of the fine gravel and sand, the percentage of which in the soil can thus be calculated.

(*b*) *Microscopical Examination.*—The residual sand should be examined with a powerful lens, or under the microscope. The quartz particles, $SiO_2$, show a crystalline structure and are generally transparent, though possibly coloured with a little ferric oxide.

Flakes of mica are common and may usually be detected, whilst rounded fragments of coloured opaque material may consist of more or less altered felspar, oxide of iron, or other minerals.

### **39.** *Clay.*

(*a*) The various washings from determination of the sand (**38** *a*) are united, and the liquid just made acid by the addition of hydrochloric acid. The sediment which settles out on standing consists of "clay." The clear liquid is poured off, the sediment washed into a weighed evaporating dish, dried on the water-bath, ignited to a dull red heat, and weighed. This gives the percentage of clay.

(*b*) *Hygroscopic Moisture and Water of Hydration.*—The hygroscopic moisture in a sample of clay is determined in the usual manner by drying about 2 grams in the steam oven to constant weight.

The chemically combined water is estimated by

igniting the residue from the hygroscopic moisture to dull red heat, and observing the loss in weight.

(*c*) A small quantity of grey, green, or blue clay, from the deep-seated formations, may be observed, on ignition, to turn red. This is caused by the iron, present generally in ferrous silicates like glauconite, becoming oxidised to the ferric condition.

Chemically speaking, clay is mainly kaolinite, $Al_2O_3.2SiO_2.2H_2O$, but most clays on closer examination will be found to contain very fine grains of quartz, felspar, calcium carbonate, etc.

*Classification of the Soils, with Characteristic Composition.*

| | Sand. | Clay. | Limestone. | Loss on Ignition. |
|---|---|---|---|---|
| Sand | 80 % | 12 % or under | ... | 3 % |
| Loam | ... | 40-70 % | less than 3 % | 3-7 % |
| Marl | ... | ... | 5-20 % | ... |
| Calcareous | ... | ... | over 20 % | ... |
| Clay | ... | ... | less than 1 % | 8-9 % |
| Peaty | ... | ... | ... | 25 % |

## 40. Plant Nutrients in the Soil.

A fertile soil must contain reasonable quantities of nitrogen, phosphoric acid, and potash, besides lime, and small quantities of iron and magnesia. Each may be tested for qualitatively :—

(*a*) *Phosphates.*—A few grams of soil are ignited in a basin or crucible, allowed to cool, and boiled in a small flask with about 10 c.c. of strong nitric acid. After cooling, a little water is added, together with some ammonia, the latter in insufficient quantity to neutralise all the acid. The liquid is then tested for $P_2O_5$ with ammonium molybdate solution.

(*b*) *Potash.*—About 5 grams of the soil are boiled with 30 c.c. of dilute HCl for five to ten minutes.

The liquid is filtered into a porcelain dish, the contents evaporated to dryness, and the residue ignited. When cold, the residue is extracted with cold water and again filtered. The clear filtrate is tested for potassium—

(i.) By adding acetic acid and then some sodium cobaltinitrite solution. A reddish precipitate indicates potash.

(ii.) By addition of platinic chloride solution (see p. 10).

(*c*) *Nitrogen.*—In normal soils nitrogen will be present chiefly in the organic form, but a small proportion should also exist as nitrates (see **41**, p. 65).

*Organic Nitrogen.*—A rough idea of the amount and nature of the organic nitrogen of a soil may be obtained by heating about 2 grams of the sample with an equal weight of soda-lime[1] in a test-tube. The amount of ammonia liberated should be compared with that from a soil of known composition, and should be confirmed by measuring the loss in weight on ignition. The latter figure is generally roughly proportional to the organic nitrogenous matter present.

*Nitrates.*—Twenty to 30 grams of the soil are extracted by shaking for five minutes with 100 c.c. of distilled water. The soil is allowed to settle, the clear liquid filtered off and evaporated to a small bulk. The residue is tested for nitrates by adding pure strong sulphuric acid, and then a drop or two of diphenylamine dissolved in sulphuric acid. A blue colour indicates the presence of nitrates.

The sulphuric acid used must be perfectly pure, since the ordinary acid generally contains sufficient nitric acid rapidly to produce a blue colour with diphenylamine.

[1] Soda-lime is lime slaked with caustic soda solution instead of water.

(*d*) *Iron.*—A few grams of the soil should be extracted by boiling for ten minutes with strong hydrochloric acid. An equal bulk of distilled water is added, the insoluble material filtered off, and a portion of the filtrate tested direct for iron by the addition of potassium ferrocyanide solution.

Most soils contain quite sufficient iron for the requirements of the crops grown on them.

(*e*) *Lime and Magnesia.*—The quantity and value of lime in a soil is usually estimated by the amount of carbonate present, and this may be approximately estimated by the method described above (**36** *b*, p. 57).

The magnesia may be detected by taking a portion of the hydrochloric acid extract from (*d*), adding ammonia and ammonium chloride to remove iron, alumina, etc., and then ammonium oxalate to precipitate the calcium. The filtrate from these operations is evaporated to small bulk, mixed with strong ammonia and sodium-phosphate solution and allowed to stand. Magnesia is thus precipitated as magnesium ammonium phosphate, $MgNH_4PO_4$.

# CHAPTER VI

## Chemical Properties of Soil

### Nitrification.

THROUGH the agency of bacteria, nitrogenous compounds in the soil are eventually oxidised to nitrates, in which form they are capable of being assimilated by growing plants. The process of nitrification goes on in two stages: one set of bacteria produce *nitrites*, which are then oxidised further to *nitrates* by micro-organisms of the type of *nitrobacter*. A base, such as calcium or magnesium carbonate, must be present in the soil, because nitrification ceases as soon as the medium becomes at all acid.

**41.** *Nitrification of Ammonium Compounds.*

A nutrient solution is made up containing the following materials per 1000 c.c.:—

| | |
|---|---|
| Potassium phosphate . . . | 5 grams |
| Magnesium sulphate . . . | 2 " |
| Ammonium sulphate . . . | 2 " |
| Common salt. . . . . . | a trace |

100 c.c. of this solution are placed in two small flasks, which are closed with plugs of cotton-wool, and the contents then boiled gently for fifteen minutes. The flasks and contents are allowed to cool, the plugs removed, and to each is added about ·5 gram of calcium carbonate, free from nitrates, which has been previously sterilised by heating. To one flask only is also added about one-fifth of a gram of ordinary arable soil. The two flasks, one containing the original solution and the other inoculated with active soil, are again closed with the cotton-wool plugs and placed in a dark cupboard in a warm room. (The optimum temperature of nitrification is 37° C.)

Every three or four days 5 c.c. of solution are removed from each flask and tested for nitrites by means of naphthylamine and sulphanilic acid (see **217**, p. 224), and for nitrates with diphenylamine and sulphuric acid (**40** (*c*), p. 63). By this means the course of nitrification can be readily traced.

The experiment can be rendered quantitative by estimating the nitrites and nitrates by the methods described under Water Analysis (p. 223), or in **48** (*d*), p. 70. If this is done the results should be plotted graphically.

**42.** *Effect of Acidity.*

If the above experiment be repeated with the omission of the calcium carbonate and the addition instead of just sufficient dilute sulphuric acid to render the solution faintly acid, the effect of acidity on nitrification will be plainly seen.

**43.** *General Occurrence of Nitrifying Bacteria.*

(*a*) *In Soils.*—Several samples of soils should be tested for nitrates according to the method previously described (**40** (*c*), p. 63). The presence of nitrates indicates that the soil is capable of nitrifying.

(*b*) *In Dust.*—Laboratory dust will generally respond to tests for nitrites and nitrates, owing to the general distribution of the nitrifying micro-organisms. If the test is negative or equivocal, the existence of the bacteria should be confirmed by seeding some of the culture solution with a small quantity of the dust, under the conditions detailed above.

## Denitrification.

In the presence of a large excess of organic matter, nitrates and nitrites are reduced by bacterial action, with the formation of gaseous nitrogen. This is termed *denitrification*, although the term is also applied to the formation of nitrogen from ammonia and nitrogenous organic matter. The bacterial activity may be shown as follows :—

**44.** *Measurement of Denitrification.*

A litre flask is nearly filled with fresh horse dung or manure, together with 1 or 2 grams of sodium or potassium nitrate. The vessel is then completely filled with water and closed with a rubber stopper fitted with a delivery tube which passes under the end of a long graduated tube filled with caustic-soda

(*b*) *Nitrate.*—In order to compare the amount of nitrate present in the original solution and that shaken with soil, 5 c.c. of each liquid is evaporated nearly to dryness, mixed with excess of *pure* strong sulphuric acid and a drop of diphenylamine dissolved in sulphuric acid. The depth of blue coloration produced in each case may be used as an indication of the nitrate present.

(*c*) *Potash.*—Equal quantities of the two solutions are acidified with acetic acid and tested with a fresh solution of sodium cobaltinitrite, and the extent of precipitation in each case noted.

(*d*) *Phosphate.*—The amounts of phosphate present in the original solution and in the soil filtrate are compared by precipitating each with ammonium-molybdate solution after acidification with a few drops of dilute nitric acid.

**48.** *Absorption of Salts (Quantitative).*

Into each of four stoppered bottles is placed 50 grams of good clay soil. To (1) is added 250 c.c. of a 1 per cent. solution of ammonium sulphate; to (2), 250 c.c. of 1 per cent. sodium phosphate solution; to (3), 250 c.c. of 0·5 per cent. solution of potassium chloride; and to (4), 250 c.c. of 1 per cent. sodium nitrate. The bottles are then well stoppered, put on one side for a week, and shaken occasionally.

In the meantime the strengths of the various solutions are estimated as follows:—

(*a*) *Ammonium Sulphate.*—Two hundred and fifty c.c. of the solution are mixed with excess of caustic soda, and distilled in the Kjeldahl distillation apparatus, the distillate being collected in 20 c.c. of $\frac{N}{2}$ acid, tinted with methyl orange. After half an hour the

distillation is stopped, and $\frac{N}{2}$ caustic soda is run into the distillate until it just turns yellow. Each cubic centimetre of $\frac{N}{2}$ sulphuric acid used up during the distillation is equivalent to 0·007 gram of nitrogen. From this may be calculated the nitrogen in 250 c.c. of the original solution.

(*b*) *Sodium Phosphate.*—Fifty c.c. of the solution are mixed with a quantity (about 1 gram) of ammonium nitrate, and treated with 20 c.c. of ammonium-molybdate solution. The mixture is stirred well, and allowed to stand in a warm place for a day. The precipitate is then carefully washed once or twice by decantation, completely transferred to a porcelain evaporating basin, and taken down to dryness on the water-bath. The residue is carefully heated until all the ammonia is given off and the substance assumes a uniform dark blue colour. Each part of the residue contains 0·0396 part of $P_2O_5$. From this the content of $P_2O_5$ in the solution may be calculated.

(*c*) *Potassium Chloride.*—Fifty c.c. of the potash solution are acidulated with a little concentrated hydrochloric acid. A strong solution of platinic chloride (10 per cent.) is added in moderate excess, and the liquid evaporated on the water-bath to a thick paste. The residue is digested with 80 per cent. alcohol, then washed by decantation with the same liquid, and finally transferred to a tared filter, where washing with diluted alcohol is continued until the washings are colourless. The residual precipitate is dried to constant weight at 100° C. It consists of $K_2PtCl_6$, and the weight multiplied by 0·1937 gives the weight of potash, $K_2O$, in the solution taken.

(*a*) *Sodium Nitrate.*—One hundred c.c. of the

solution are poured into a beaker containing a zinc-copper couple reaching above the level of the liquid. The solution is acidified with dilute acetic acid and left overnight in a warm place, the beaker being covered with a clock-glass. The nitrate is thus reduced to ammonia, which is estimated by pouring the liquid, together with the washings of the couple, into a distillation flask, making alkaline with caustic soda and proceeding as described above under Ammonium Sulphate (48, *a*).

*Zinc-Copper Couple.*—This is prepared by immersing a few pieces of zinc foil in a dilute solution of copper sulphate until they are just covered with a firm deposit of precipitated copper. The "couple" thus obtained is washed with distilled water and kept for use in a wide-necked bottle of distilled water.

The solutions which have been standing in contact with the soil are now filtered, and the ammonia, phosphoric acid, potash, and nitrate estimated in the respective filtrates. The amount of each which has been absorbed may thus be calculated.

Ammonia and potash are absorbed very readily by soil, and phosphoric acid is also taken up to a very large extent, but in the case of the nitrates the action is very slight, and practically the whole of the nitrate of soda may be washed out from the soil by water.

In the case of potash and ammonium salts the absorption is largely effected by interaction with calcium compounds present in the soil, calcium sulphate or chloride appearing in the drainage water. This may be shown by experiment.

**49.** *Removal of Lime from the Soil.*

(*a*) If the soil used in **48** contains an appreciable quantity of lime, the filtrates from treatment with ammonium sulphate and potassium chloride are tested for calcium. A few cubic centimetres of each are rendered alkaline with ammonia and mixed with ammonium oxalate solution. A white powdery precipitate indicates calcium, the presence of which is confirmed by evaporating a little of each solution to dryness, moistening with hydrochloric acid, and testing on a platinum wire in the Bunsen burner. Calcium compounds give a yellowish red flame colour.

Limestone is also removed from the soil by the action of carbonic acid, which produces soluble calcium bicarbonate:

$$CaCO_3 + CO_2 + H_2O = Ca(HCO_3)_2$$

(*b*) About 50 grams of a calcareous soil are mixed with water in a beaker, and carbon dioxide gas bubbled into the mixture for ten minutes. The soil is filtered off, and the filtrate boiled vigorously for a few minutes. A white precipitate or turbidity indicates the presence of the bicarbonate, which is decomposed by the action of heat into normal calcium carbonate again:

$$Ca(HCO_3)_2 = CaCO_3 + CO_2 + H_2O.$$

If no precipitate is obtained (indicating only a small quantity of limestone in the soil), the filtrate should be evaporated to dryness, taken up in a little dilute hydrochloric acid, and tested for calcium as in (*a*) above.

# CHAPTER VII

## PHYSICAL PROPERTIES OF SOIL

### Colour of Soil.

THE colour of any soil under examination should be noted, since within certain limits it is an indication of fertility, black and red soils being the most fertile, *e.g.* the black wheat-growing soils of Russia, and the red soils of Devon and Dunbar. The colour is largely dependent on the proportion of organic matter and iron compounds.

### 50. *Observation of Colour.*

Soil samples to be compared should be in the air-dry condition. The colour is noted and the sample gradually heated to dull redness. During heating the following points should be noted :—

(*a*) The colour at 100° C., *i.e.* after removal of the moisture.

(*b*) The extent of carbonisation, and whether a nitrogenous odour is produced.

(*c*) Any change of colour which may ensue.

(*d*) The colour of the residue both in the dry condition and after moistening with water.

Soils may be roughly classified according to colour, as follows :—

| | |
|---|---|
| Black or dark brown when moist . . . | High humus content. |
| Red or reddish brown | Ferric oxide, in a particular state of division. |
| Bluish tint, changing to a reddish shade on heating | Ferrous compounds and organic matter. |
| Light yellow or whitish tint . . . . | Preponderance of sand, or else chalk. |

### Specific Gravity.

Since soil consists of a number of particles which enclose a certain amount of air or water, it will have a "true" and an "apparent" density. The former represents the weight per unit volume of the solid particles entirely free from air, whilst the latter is that of the soil with pore-spaces included, and is, of course, considerably smaller than the true specific gravity. A soil in good tilth, for example, will tend to have a low apparent specific gravity. In general the specific gravity of a soil decreases inversely to its humus content. The true density of ordinary soils may be taken as 2·65 ; the apparent density varies, but averages approximately 1·2.

51. *Determination of True Specific Gravity.*

The specific gravity is generally determined at 20° C. in an ordinary specific gravity bottle.

A 100 c.c. capacity specific gravity bottle is weighed, then filled with freshly boiled distilled water and weighed again. Ten grams of the soil dried at 100° C. are boiled for a short time in a beaker with a few c.c. of water in order to expel the air. The soil is then completely washed into the gravity bottle with boiled distilled water, cooled to 20°, and filled to the mark with the water. The bottle is then weighed again.

The weight of soil used divided by the weight of water displaced gives the specific gravity of the soil.

52. *Determination of Apparent Specific Gravity.*

In order to ascertain the apparent density of a soil the best method is to get a smooth open measure, such as a pint pot or a specimen jar, fill it with the soil, with gentle tapping, strike off the upper surface flush with the top of the pot, and weigh. The volume of the vessel is measured by filling it with water and the apparent density obtained by dividing the weight of soil by the volume.

**Pore-space.**

53. *Direct determination in Sand.*

Fifty grams of sand are weighed out and transferred to a dry measuring cylinder. The latter is tapped gently on the bench and the volume occupied by the sand noted. (The apparent density may be measured from this datum.) The sand is now removed and 50 c.c. of water poured into the cylinder, after which the sand is carefully returned to the vessel, the mixture stirred or shaken to remove any occluded air, and the volume again noted. From these observations the percentage of air-spaces in the sand may be readily calculated.

**54.** *Pore-space of Soils by Calculation.*

The pore-space of a soil may be readily calculated from a knowledge of its true and apparent densities. The experiment should be carried out with samples, taken *in situ*, of (*a*) a clay soil, (*b*) a light sandy soil, and (*c*) a calcareous soil.

The apparent density may be determined as above (**52**), or may be obtained by weighing the samples taken in the field by the cubic sampler (see p. 84), of known dimensions. The true specific gravity is found on the soil dried at 100° C., as described in **51**.

If $D$ = real specific gravity,
$D_1$ = apparent specific gravity.

Then the weight of soil equal to the pore-space of unit volume $= D - D_1$.

$$\therefore \text{Volume of pore-space in unit volume} = \frac{D - D_1}{D}$$

*i.e.*, *Percentage of pore-space in soil* $= \dfrac{100\,(D - D_1)}{D}$

## Absorption of Water by Soils.

The retention and rate of percolation of water through a soil is very largely dependent on its texture, and the amount of organic matter which is present.

**55.** *Hygroscopic Moisture.*

The percentage of water present in air-dried peat, sand, and clay is determined by drying weighed quantities of material (about 10 grams) to constant weight in the steam oven. The loss in weight, representing the hygroscopic water, is highest in peat (humus) and lowest in sand.

### **56.** *Retention of Water by Soils.*

The amount of water which a soil can hold is measured by the pore-space plus the quantity of water capable of being absorbed by the particles without disturbing their condition. The estimation of the water-holding capacity of a soil is of importance.

*Apparatus.*—The apparatus consists of a small brass cylinder 6 cm. in diameter and 10 cm. long, having a perforated bottom. The capacity of the cylinder is about 280 c.c., and is determined accurately by placing while slightly hot on a slab of wax so as to fill up the holes, and then running in water from a burette until full.

*Procedure.*—A disk of moist muslin is placed over the perforated bottom, any excessive moisture removed, and the cylinder weighed. It is then carefully filled with the "fine earth" of the soil (**64**, p. 84) under examination, the sample being compacted by gently tapping on the bench. The surface is then struck off level and the cylinder again weighed, this giving the weight of soil taken. The cylinder is then placed in a vessel of distilled water, the surface of the water being kept just below the top of the cylinder. The whole apparatus is covered over and allowed to stand for one hour, or until free moisture appears on the surface of the soil, thus showing that the maximum amount of water has been absorbed. The cylinder is then removed, wiped dry, and again weighed. The increase of weight gives the amount of water absorbed.

The moisture present in the "fine earth" must be determined if this has not already been done.

*Calculation of Results.*—The water-holding capacity of the soil should be stated as percentage volume of water contained in the soil. Thus, if

$w$ = Weight of moisture in soil,
W = Weight of water absorbed,
V = Volume of soil.

Then, water-holding capacity

$$= \frac{(w+W)\,100}{V}$$

The experiment should be carried out with sand, with fine earth from a loam and from a clay soil, and also with a peaty soil or sand mixed with organic matter. The results should be compared with the air-space, determined as in **54**.

### **57.** *Percolation of Water in Soils.*

A small percolator (Fig. 13) is fitted at the bottom with a piece of wire gauze, and is then filled up to a definite mark, about 2 inches from the top, with the soil under examination, the sample being compacted by gentle tapping. Water is then added, and a constant level is maintained by inverting a flask filled with water at the required height, as in diagram.

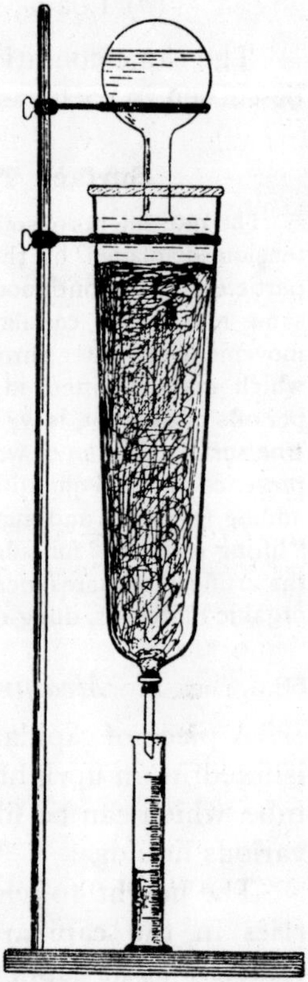

FIG. 13.—Apparatus for measuring the rate of percolation of water through soil.

The time taken for the water to reach the bottom, and the number of cubic centimetres percolating per hour should be determined, and tabulated for comparison with the following soils:—

(*a*) Sand. (*c*) Clay soil.
(*b*) Loam. (*d*) Peaty soil.

The same conditions of experiment must, of course, be chosen in each case.

## Surface Tension and Capillarity.

The movements of soil water are largely governed by the surface tension displayed by the liquid when in contact with the soil particles. The continuous pore-spaces of the soil, acting in the same manner as capillary tubes, allow the upward and lateral movement of water through the soil, and the quantity of water which may be lifted in this way from the subsoil, *e.g.* during periods of drought, is very considerable and of great importance. The surface tension of water is high, but is readily affected by the presence of small quantities of material in solution; the effect of adding fertilisers and manures to the soil consequently alters the "lifting capacity" for subsoil water. Most of the salts present in the artificial manures increase the surface tension of water, whereas organic materials, dung extract, etc., have a contrary effect.

### 58. *Measurement of Capillarity.*

A piece of capillary tubing about 4 inches in length is fixed in an upright position in a small glass vessel or tube, which can be filled exactly to the same height by various liquids.

The height to which the liquid under examination rises in the capillary tube is measured by a cathetometer. The capillarity of the following liquids should be measured:—

(1) Distilled water.
(2) Alcohol.
(3) Dung liquor.
(4) Sulphate of ammonia solutions, 1, 5, 10 and 20 per cent.

(5) Nitrate of soda solutions of similar strengths.

(6) Soil extract.

*Soil extract.*—The soil extract is prepared as follows: Ten grams of the soil are rubbed up with 15 to 20 c.c. of distilled water, and allowed to stand for twenty-four hours with frequent stirring. Any fine particles not removable by filtering may be neglected, even if the solution is turbid. The radius of the capillary tube is obtained from the weight of a thread of mercury filling a measured length. If $l$ = length of mercury thread, and $w$ = weight of mercury, then $r$, the radius of the tube,

$$= \sqrt{\frac{w}{13 \cdot 59 \times \pi l.}}$$

The surface tension (T) of the various solutions is obtained approximately from the formula—

$$T = \frac{hdr}{2}$$

where $h$ is the height of the liquid in the tube in centimetres and $d$ is the density of the solution. The result is obtained in *dynes*.

## 59. *Capillary Attraction in Soils.*

Long glass tubes of 1-2 cm. in diameter and graduated in centimetres, should be used for this determination. The lower end of each tube is closed with a piece of linen, and the "fine earth" of the soils under examination filled in little by little with gentle tapping on the bench for compacting.

The tubes, after filling, are supported in an upright position in a vessel containing water (Fig. 14), the linen-covered ends dipping to the depth of 2 cm. The height of the water is measured at stated intervals, and the results for the various soils plotted out graphically. The observations may be discontinued after one hundred and twenty hours, at the end of which time the tubes may, if desired, be cut into pieces 10 cm. in length and the percentage of water determined in each portion.

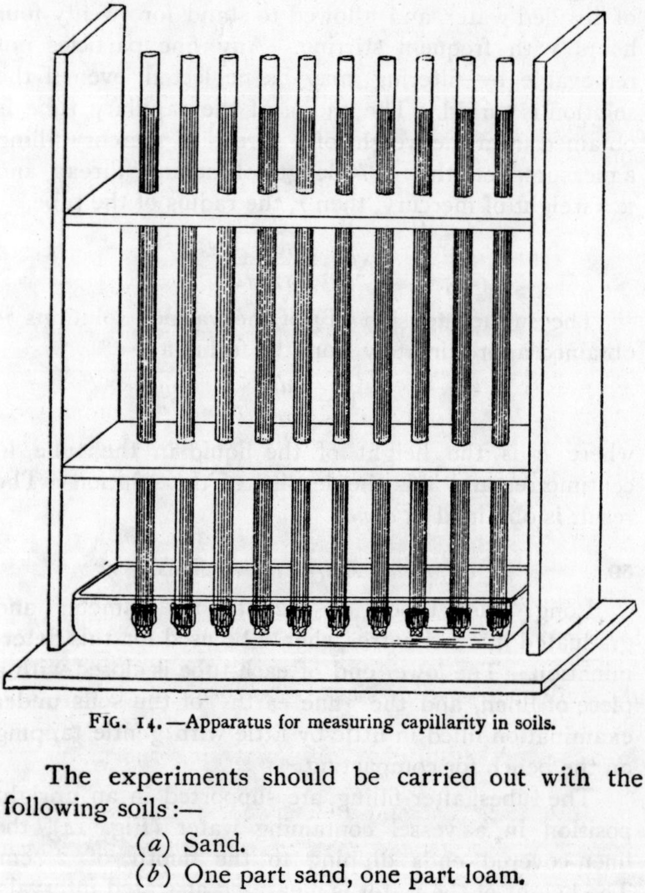

FIG. 14.—Apparatus for measuring capillarity in soils.

The experiments should be carried out with the following soils :—

(*a*) Sand.
(*b*) One part sand, one part loam.
(*c*) Loam.
(*d*) Clay soil.

### Soil Thermometry.

The measurement of soil-temperature at different depths is sometimes resorted to in connection with the characterisation of different soils. Field experiments on the subject require special attention and special forms of thermometers, but the rate of absorption of heat by a soil may be measured and compared with a standard soil by means of a simple apparatus.

**60.** *Absorption of Heat by Soil.*

A cubical metal box about 4 inches square is filled compactly with the "air-dried fine earth" (**64**) of the soil to be examined. The box should be covered with felt to prevent access of heat except at the exposed surface, and should be provided with small openings at different depths to allow of the insertion of thermometers. The open end is then exposed for a few hours to the direct rays of the sun, and the temperature at various depths noted at intervals.

The rate of heating should be compared, in the cases of a sandy and a heavy soil, both in the dry and moist condition, and the results plotted out graphically. Dark and light coloured soils should also be compared. The temperature of the surrounding air must be noted, and the conditions, as far as possible, chosen identically in all the experiments. For this reason it is as well to have several similar boxes in use at the same time.

### Flocculation and Deflocculation of Clay.

The very fine particles which compose that fraction of soil known as clay, may, in certain circumstances, be aggregated or coagulated to form larger granules, which themselves act as separate soil particles. This phenomenon, termed ***flocculation***, is most important from the point of view of cultivation of clay soils, since it is highly desirable to obtain the soil in a coarse-grained condition capable of being worked. If the clay has undergone ***deflocculation*** or resolution into its finest particles it becomes sticky, persistently wet, and "puddles" when worked. The desirable flocculation or granulation of the clay particles is aided by certain salts and acids, and hindered, sometimes for an indefinite period, by alkalies.

### 61. *Effect of Acids, Alkalies, Salts, etc.*

A few grams of clay are rubbed up with distilled water in a mortar, the thin paste diluted to 1 litre, allowed to stand for a short time, and the turbid liquid poured into a number of cylinders (100 c.c. measuring cylinders or Nessler tubes do excellently); one of these cylinders is used as a control, and to the others are added the following materials:—

(*a*) 1 c.c. $\frac{N}{2}$ Hydrochloric acid.

(*b*) 0·5 c.c. $\frac{N}{2}$ Hydrochloric acid.

(*c*) 1 c.c. 1 per cent. Sodium chloride solution.

(*d*) 1 c.c. 1 per cent. Calcium chloride solution.

(*e*) 4·5 c.c. Lime water.

(*f*) 0·5 c.c. $\frac{N}{2}$ Ammonium hydroxide.

(*g*) 1 c.c. 1 per cent. Sodium phosphate solution.

The respective times taken for the columns of liquid to become clear, *i.e.* for the clay to flocculate, should be noticed, and the materials classified according to their activity. By varying the quantities of salts used, the relationship between the amount of salt present and the extent of flocculation may be determined.

The flocculating effect of lime in a soil is due to its conversion into carbonate and then into bicarbonate. Solutions of the latter salt have a very aggregating effect.

62. *Shrinkage of Clay on Drying.*

A small brick of modelling clay is made, of 6 inches in length. Two marks, exactly 5 inches apart, are made on the clay, and the brick weighed. As the brick dries the distance apart of the two marks is measured from time to time, and the loss in weight determined after each measurement. Eventually drying is completed in the oven, a final measurement and weighing made, and curves drawn to show the connection between shrinkage and water-content.

The experiment should be repeated with a brick composed of clay mixed with about $\frac{1}{2}$ per cent. of quick-lime, and the results compared with those previously obtained.

# CHAPTER VIII

## MECHANICAL ANALYSIS OF SOIL

THE mechanical analysis of a soil consists of its division into various grades of particles according to size. This is a most important point with regard to the physical properties of a soil, especially as a measure of the very fine particles present. The latter give the soil power to retain moisture and resist drought, whilst shrinkage on drying and tenacity when dry are largely dependent on low proportions of humus, sand, and chalk.

Two types of method are open for the mechanical soil-analysis, viz.: (*a*) Those depending on sedimentation in still water and decantation after a time; (*b*) Hydraulic methods, in which separation is effected by water-currents of varying speeds (elutriation). The former method is the one generally adopted in this country, although elutriation offers a great deal of scope in its application.

63. *Sampling Soil for Analysis.*

Soil samples are generally taken to a depth where soil begins to merge in subsoil, or, if this is not a shallow layer, to a depth of 9 inches, as representing the soil proper. Three methods are in general use, the surface vegetation being first removed in each case.

(*a*) A surface 6 inches square is marked out with a spade, and the sides of the square cut down to a depth of 9 inches. The surrounding soil is cut away and the block removed with the spade, either at the 9-inch depth or at the line where the subsoil commences.

(*b*) A more accurate way is to use a steel box without top or bottom, 6 inches square in section and 9 inches deep. The box is placed in position and hammered down until the top is flush with the surrounding soil. The contents are then carefully removed and conveyed to the laboratory.

(*c*) Samples are most conveniently taken by means of an auger, many types of which exist. The auger is forced gently into the soil to the required depth, the tool removed and the core extracted. The advantage of this method is that the boring may be continued into the subsoil to any depth if necessary.

### 64. *Preparation of the Sample.*

The well-mixed borings are spread out on shallow trays to dry. This may be expedited by heating in the oven to a temperature not exceeding 40° C. About 1 lb. of dry material is weighed out and then transferred to a sieve with 3 mm. diameter round meshes, and separated as much as possible. The portion which does not pass the sieve is rubbed in a mortar with a wooden pestle, care being taken not to crush the stones, pieces of chalk, etc., and the material again sieved. The portion remaining upon the sieve is well washed in a stream of water until all the fine soil is removed, any portions of vegetable matter are picked out, and the residue weighed as "stones."

The material passing the sieve is again exposed to the atmosphere for some time, and is stored in a bottle for further examination, and labelled "air-dry fine earth."

**65.** *Mechanical Analysis by Sedimentation.*[1]

(*a*) Ten grams of the air-dry fine earth are treated in a beaker with 100 c.c. of $\frac{N}{5}$ HCl, worked with a glass rod fitted with a small rubber stopper so as to break up any lumps, and allowed to stand for one hour. This dissolves the carbonates and breaks up the humates, and resolves the soil into its ultimate particles.

The soil is filtered on a tared filter paper and washed with distilled water until the filtrate is quite free from acid, after which it is dried and weighed. The loss suffered by the soil represents soluble material and hygroscopic moisture. The soil is then carefully washed on to a small sieve of 100 meshes to the linear inch with ammoniacal water, prepared by dissolving 1 c.c. of strong ammonia in half a litre of water. The muddy liquid passing through the sieve is collected in a beaker marked at a distance of 8·5 cm. from the bottom. The portion remaining on the sieve is dried and weighed, and separated into two portions by means of a sieve with 1 mm. meshes, the coarser portion being designated "*fine gravel*," and that passing through the sieve as "*coarse sand*." Further separation is effected by subsidence.

(*b*) The muddy liquid from (*a*) is well triturated with the rubber pestle, and ammoniacal water added up to

[1] Official method of the Agricultural Education Association.

the 8·5 cm. mark. The whole is then left to stand for twenty-four hours. The effect of the ammonia is to keep the clay particles in suspension, the heavier portions settling to the bottom. The turbid supernatant liquid is now rapidly poured off into a large jar or labelled Winchester bottle, and the residue again stirred with ammonia, water added to the 8·5 cm. mark, and allowed to stand for twenty-four hours. The operation is repeated until the liquid pours off quite clear, all the washings being kept together. The united washings are well shaken, their total volume measured, and 100 c.c. extracted and evaporated to dryness in a weighed dish; the residue is ignited over a Bunsen burner, cooled, and weighed. From the weight, calculated back on the whole of the washings, is obtained first the weight, and then the percentage, of "*clay*" in the soil.

(*c*) The sediment left after removal of the "clay" is again stirred well with ammonia water, which is now added to a depth of 7·5 cm. The contents of the beaker are allowed to stand for twelve and a half minutes, and the liquid poured off into a Winchester as before. This operation is repeated until the washings are quite clear, after twelve and a half minutes' sedimentation. The contents of the liquid are determined by evaporation and ignition, as before, and the quantity gives the proportion of "*fine silt.*"

(*d*) The residue from (*c*) is again treated with ammonia, as in the previous operations, the liquid being filled up to the 10 cm. mark and the time of sedimentation reduced to one hundred seconds. When the liquid comes off perfectly clear the residue is washed into a porcelain dish, dried and ignited, and weighed as "*fine sand.*" The washings from this give on evaporation and ignition the fraction known as "*silt.*"

## *Scheme for the Mechanical Analysis of Soil.*

<table>
<tr><td colspan="8">Air-dried Soil.<br>Weigh out about 1 lb. of the air-dried soil. Work in mortar with wooden pestle until no more passes through the 3 mm. sieve.</td></tr>
<tr><td rowspan="6">Residue.<br>Wash free from earth, dry and weigh. This fraction consists of</td><td colspan="7">Portion passing through Sieve.<br>This is the "air-dry" fine earth. Treat 10 grams with 100 c.c. of $N/5$ HCl. Allow to stand 1 hour, with occasional stirring. Filter and wash free from acid. Dry and weigh.</td></tr>
<tr><td rowspan="5">Filtrate.<br>Contains the dissolved matter. The loss suffered by the soil during treatment represents</td><td colspan="6">Residue.<br>Wash the residue with ammonia water on to a small sieve (100 meshes per linear inch). Collect the liquid in a beaker marked on the side at 10 cm., 8·5 cm., and 7·5 cm., from the bottom.</td></tr>
<tr><td colspan="2">Residue.<br>Separate by means of the 1 mm. sieve.</td><td colspan="4">Muddy Liquid.<br>Make up to the 8·5 cm. mark with ammonia water. Allow to stand for 24 hours, and pour off the liquid. Repeat until the washings are clear.</td></tr>
<tr><td rowspan="3">Residue.<br>Dry and weigh. Consists of</td><td rowspan="3">Portion passing through Sieve.<br>Dry and weigh. Consists of</td><td colspan="3">Residue.<br>Make up to 7·5 cm. mark with ammonia water. Allow to stand 12½ minutes, and pour off the liquid. Repeat until washings are clear.</td><td rowspan="3">Washings.<br>Evaporate to dryness, ignite and weigh. Fraction is</td></tr>
<tr><td colspan="2">Residue.<br>Make up to 10 cm. mark with ammonia water, and allow to settle for 100 seconds. Repeat as before.</td><td rowspan="2">Washings.<br>Evaporate to dryness, ignite and weigh. Fraction is</td></tr>
<tr><td>Residue.<br>Wash into porcelain dish. Dry, ignite, and weigh. Fraction is</td><td>Washings.<br>Evaporate to dryness, ignite and weigh. Fraction is</td></tr>
<tr><td>Stones and Gravel.</td><td>Soluble Matter and Hygroscopic Moisture.</td><td>Fine Gravel.</td><td>Coarse Sand.</td><td>Fine Sand.</td><td>Silt.</td><td>Fine Silt.</td><td>"Clay."</td></tr>
</table>

The size of the various fractions, expressed as diameters in millimetres, is as follows:—

| | | |
|---|---|---|
| 1. | Stones and gravel | from 3 mm. upwards. |
| 2. | Fine gravel . . | 1 to 3 mm. |
| 3. | Coarse sand . | 0·2 to 1 mm. |
| 4. | Fine sand . . | 0·04 to 0·2 mm. |
| 5. | Silt . . . | 0·01 to 0·04 mm. |
| 6. | Fine silt . . | 0·002 to 0·01 mm. |
| 7. | "Clay" . . | less than 0·002 mm. |

A résumé of the operations involved in the mechanical analysis of soil by the process of sedimentation in water is given on page 87.

67. *Mechanical Analysis by the Hydraulic Method.*

The principle of this method depends on the fact that particles in a liquid are carried upward, if the liquid is given an upward motion and its velocity is greater than the rate of subsidence of the particles in the still liquid. Within limits this is proportional to the size of the particles, in the case of soil. Various methods have been adopted for obtaining definite speeds of flow corresponding to the limiting sizes of the various grades of particles. The most satisfactory apparatus in use is probably Mayer's modification of Schöne's elutriator.

*Apparatus.*—The apparatus is shown in Fig. 15. It consists of a glass vessel fitted with a tap at the lower extremity for admitting water. For a distance upwards of 20 cm. the sides of the tube are parallel, and the bore about 1 cm. diameter. For a distance of 50 cm. the tube then expands conically until the internal diameter reaches 5 cm. For a distance of 10 cm. the vessel is again strictly cylindrical, after which it narrows rapidly until at the top it is 2 cm. in diameter. The top is closed with a stopper carrying two glass tubes, one (F) bent down so as to conduct the overflow into the receiving vessels, and the other (H), termed the "piezometer," for the purpose of regulating the flow by the height attained therein by the column of water.

The standardisation of the apparatus consists principally of altering the size of overflow-opening F, until with a piezometer pressure of 5 cm., 100 c.c. of water will pass over in one minute.

The use of the lower conical portion of the elutriator is twofold: viz. (*a*) to secure a gradual decrease in the velocity of the rising current; (*b*) to produce an agitation among the soil particles for the purpose of disintegrating any aggregates. The object of the cylindrical portion is to secure a uniform speed in the rising current. For uniform currents there is a fairly definite determinable relationship between the speed of the current and the maximum size of the particles carried out of the apparatus.

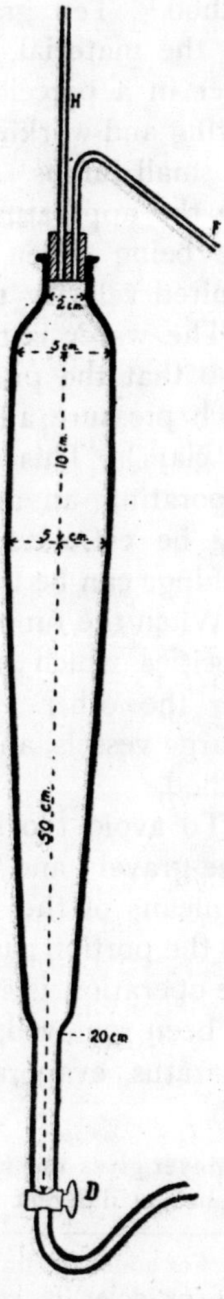


FIG. 15.—Mayer's modification of Schöne's elutriator.

68. *Procedure.*—The "air-dry fine earth," freed from stones and gravel by sifting (see above), is treated with hydrochloric acid as in the preparation of the sample for the sedimentation

method.[1] Ten grams are used for each experiment, and the material, washed free from acid, is boiled with water in a porcelain dish for one hour, with constant stirring and working with the rubber pestle to break up the small lumps. The soil, thus treated, is introduced into the apparatus, the water allowed to enter slowly, care being taken to avoid reaching even the lowest required velocity until the outflow begins.

The water is then regulated by means of the tap D, so that the pressure in the piezometer is 2 cm., at which pressure all that passes over may be regarded as "clay." This may be collected and estimated by evaporating an aliquot part to dryness; or the clay may be estimated by difference, in which case the washings can be thrown away.

When the runnings are perfectly clear, the different velocities which have been decided on are used, one after the other. The various fractions are collected in large vessels, and estimated as in the sedimentation method.

To avoid too high pressures in the piezometer, the "fine gravel" and "coarse sand" may first be separated by means of the 100 meshes to the linear inch sieve, and the portion passing through reserved for elutriation. The operation is stopped when all but the "fine sand" has been removed, and the latter is washed out of the apparatus, evaporated to dryness, and weighed forthwith.

Mayer gives the following table as representing the products obtained at different pressures, but the apparatus can obviously

---

[1] For soils containing large quantities of undecomposed limestone or dolomite particles, the use of HCl may lead to the inclusion in the silt of many particles which could not be regarded as coming from the soil itself as it exists. The HCl treatment may be recommended unconditionally for alluvial soils.

be standardised for any size of particle. To bring the elutriator into line with the settling process, the fractions from the latter may be placed in the apparatus, and the water-pressure necessary to carry them over noted.

| Fraction. | Average Diameter of Particles. | Piezometer Pressure. |
|---|---|---|
| | mm. | cm. |
| Clay | ... | under 2 |
| Very fine silt | 0·010 | 2 |
| Fine silt | 0·016 | 4 |
| Medium fine silt | 0·025 | 16 |
| Silt | 0·036 | 64 |
| Coarse silt | 0·047 | 256 |
| Very coarse silt | 0·072 | 1024 |
| Very fine sand | 0·12 | 4096 |
| Fine sand | 0·16 | > |

# CHAPTER IX

## Chemical Analysis of Soil

THE chemical analysis of a soil is carried out with a view to ascertaining what quantities of the plant nutrients are present, so that any actual deficiencies can be made good, or the amounts so altered as to meet the requirements of particular crops.

The most important constituents estimated are the carbonates (representing the "available basicity" of the soil), the organic matter, nitrogen, phosphorus, potassium, calcium, and magnesium. Of lesser importance are sodium, iron, aluminium, manganese, chlorine, and sulphur (as sulphate).

### 69. *Sampling.*

The "air-dried fine earth" passing through the 3 mm. sieve is prepared as previously described for the mechanical analysis (**64**). About 100 grams of this material are then further broken down in a steel mortar until it will all pass through a sieve with holes 1 mm. in diameter. This is the soil used for analysis.

70. *Moisture.*

About 3 grams of the powdered soil are dried in a porcelain basin in the steam oven until of constant weight, or for a maximum period of twenty-four hours. The loss in weight represents the hygroscopic water.

### Organic Matter, Nitrogen, etc.

71. *Loss on Ignition.*

The loss on ignition is intended primarily to measure the amount of organic matter present, but unavoidably includes also a quantity of the water of hydration, or water combined with certain of the soil silicates. If too high a temperature is used some of the carbonates may also be decomposed.

*Procedure.*—The residue from estimation of the moisture (**70**) is heated over a very small flame from a rose burner, or from an argand burner, preferably in a platinum dish, for several hours, with occasional stirring. The temperature should not be raised above a dull redness, and after ignition and allowing to cool, the mass is moistened with a few drops of ammonium carbonate solution. The residue is again slowly heated to about 150° C., to drive off the excess of ammonia, allowed to cool in a desiccator, and weighed.

72. *Organic Matter.*

A more accurate determination of the organic matter may be made by oxidising the soil with chromic acid and measuring the amount of carbon dioxide given off.

*Procedure.* — The apparatus described below (**75**) for the estimation of carbonates may be used, with slight alterations, for the estimation, which may follow with the same sample immediately after the determination of carbonates.

A small reflux condenser and a tube of heated copper oxide should be interposed between the combustion flask and the first absorption tube. The CuO tube is used to complete the combustion of some of the products formed; 5-10 grams of powdered potassium bichromate are added to the combustion flask A (Fig. 16); a current of air is drawn through the apparatus by means of the aspirator for a few minutes, and concentrated sulphuric acid then added by means of the dropping funnel. The contents of the flask are then heated until the acid begins to give off fumes, and the temperature maintained at this stage for about ten minutes, or until the combustion of the organic matter is complete.

In calculating the percentage of organic matter it is assumed that its carbon content is 55 per cent., so that the weight of $CO_2$ formed multiplied by the factor 0·496 gives the organic matter.

**73.** *Nitrogen.*

The nitrogen is estimated by Kjeldahl's method (see **7**). Ten to 20 grams of the powdered soil are used, and no special precautions need be taken for the nitrates existing in the soil, since their quantity is so small as to be almost negligible beside the total nitrogen present.

In the subsequent estimation of the ammonia formed it is desirable to employ steam distillation, owing to the amount of insoluble matter present, which prevents regular ebullition. Instead of steam distilling, the sulphuric acid digest may be diluted with water, filtered through a large Buchner funnel, and an aliquot portion of the filtrate distilled with NaOH in the usual manner.

**74. Total Carbonate.**

If the quantity of carbonate present in the soil is large, its amount may be determined (*a*) by loss of weight; (*b*) by absorption. The use of the Schrötter apparatus for the former method has already been described.

75. *Method of Absorption.*

The apparatus shown in Fig. 16 is employed.

The experimental flask A, of about 300 c.c. capacity, is fitted with a dropping funnel containing the dilute

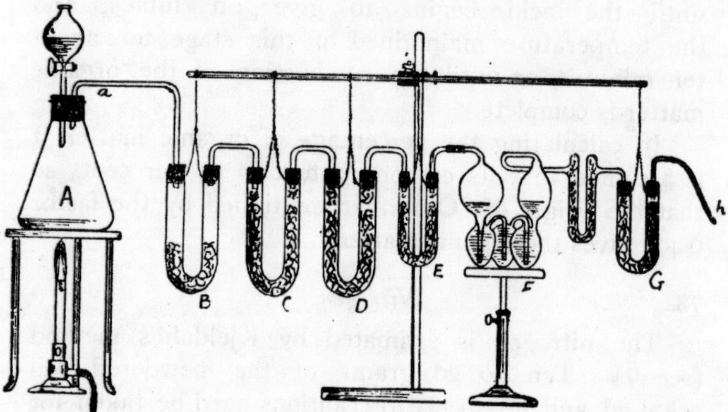


FIG. 16.—Estimation of carbonates in soil by absorption.

acid necessary for the decomposition, and with the outlet tube *a*, which is connected with a series of tubes for purifying the $CO_2$ previous to its absorption. The U-tubes B and C are packed with anhydrous calcium chloride. The tube D is filled with pumice stone, which has been soaked in a saturated solution of copper sulphate and dried at 200° to 300° C. This tube is for the purpose of absorbing any hydrochloric acid vapour which may be carried over, and may obviously be left out if sulphuric acid is employed. The small U-tube E also contains calcium chloride. The actual absorption of the carbon dioxide occurs in the potash bulbs F, which are charged with 50 per cent. solution of caustic potash, and in the small attached U-tube, which is packed with small pieces of solid potash. The use of soda-lime instead of potash is not recommended. The U-tube G prevents the absorption of carbonic acid or moisture from the atmosphere. It is filled with fused calcium chloride in the limb next to the absorption apparatus, the outer limb being packed with soda-lime.

*Preliminary operations.*—As commercial calcium chloride always contains a certain amount of quicklime, and consequently absorbs carbon dioxide, it is necessary to pass $CO_2$ through the calcium chloride U-tubes for an hour or so, in order to transform any oxide into carbonate. The excess of carbon dioxide is removed by a current of dry air.

The potash bulbs and U-tube are carefully weighed and the apparatus is then fitted together, the dropping funnel being filled with dilute sulphuric acid. Hydrochloric acid is not as convenient for use.

The apparatus is tested to see whether all the joints are tight, by sucking gently at *h*, clipping the rubber tube tightly between the fingers, and observing whether any leakage causes the bubbles to move in the potash bulbs. It is essential that the apparatus be quite tight.

*The estimation.* — According to the amount of carbonate shown to be present in the soil by a preliminary test, 5 to 20 grams of the material are introduced into the flask and made into a thin paste with a little water, the apparatus closed and again tested as to tightness. The acid is now gradually run in, the evolution of the carbon dioxide being maintained at such a rate that about two bubbles per second pass into the potash bulbs. When evolution of the gas has ceased, the remainder of the acid is run in and the flask gently heated. An aspirator is attached at *h*, the tap of the dropping funnel opened, and a gentle stream of air is drawn through the apparatus in order to displace any $CO_2$ which remains. Carbonic acid from the atmosphere is excluded by attaching a small soda-lime tube to the dropping funnel.

When all the carbonic acid has been absorbed the potash bulbs and U-tube are disconnected and weighed. The increase in weight is the amount of $CO_2$ present in the carbonate.

**76.** *Amos' Method.*

This method, which is suitable for the estimation of small quantities of carbonate (under 1 per cent.), depends on the well-known double titration principle. The carbon dioxide is liberated from the soil with acid and absorbed by caustic-soda solution. The alkali is then titrated, using successively phenolphthalein and methyl orange as indicators.

*Apparatus.*—The construction of the apparatus may be gathered from Fig. 17. It consists of:—

A. A Reiset absorption apparatus containing 100 c.c of a 4 per cent. sodium hydroxide solution, for washing the air free from carbon dioxide.

B. A Jena flask, in which is placed the soil to be treated. The flask is provided with a side-tube.

C. A second Reiset apparatus containing 100 c.c. of 4 per cent. sodium hydroxide solution for absorbing the $CO_2$ liberated in B.

D. A small reflux condenser.

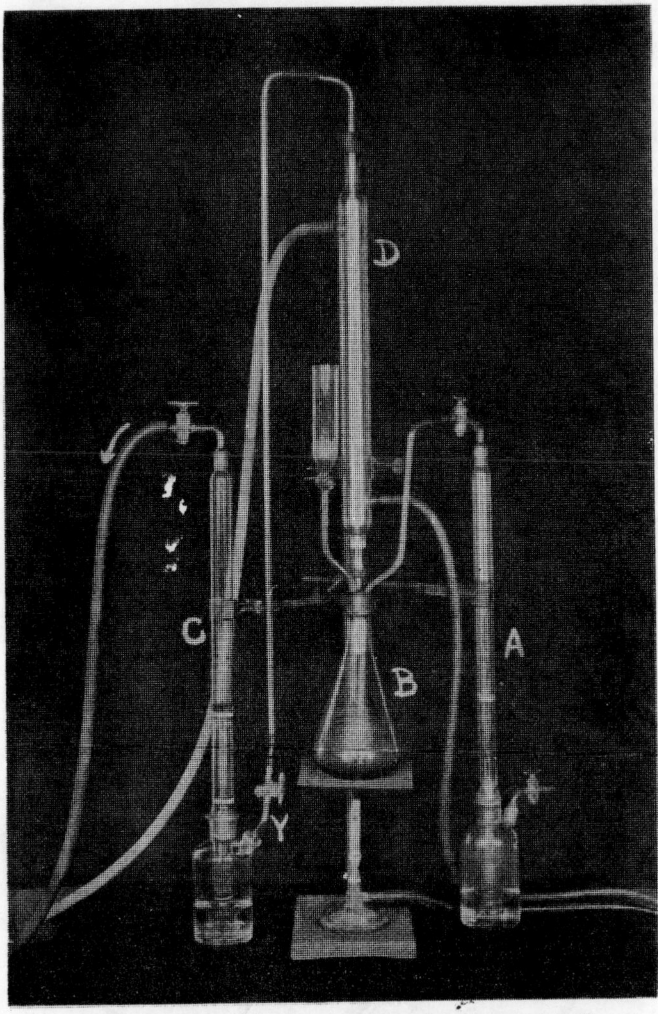

FIG. 17.—Amos apparatus for estimating carbonates in soil.

[*Face page* 96.

*Procedure.*—A quantity of the soil (not exceeding 50 grams) is weighed out, in which the amount of limestone present is not more than 0·5 gram. The material is introduced into the flask B, shaken up with 75 c.c. of carbon-dioxide-free water, and the end Y of the rubber tube connected directly with the pump, so that a stream of air is drawn through the apparatus and all the atmospheric $CO_2$ swept out. The apparatus is now connected up as in the diagram, and a steady stream of air drawn through. Twenty c.c. of strong hydrochloric acid are run into B by means of the dropping funnel, and the contents of the flask gradually brought to the boiling point; the boiling is continued for twenty minutes, to ensure all the carbon dioxide being swept into the Reiset C.

The titration is carried out in the lower part of the Reiset apparatus, into which the contents of the absorption tube are washed, phenolphthalein is added, and normal hydrochloric acid run in until the pink colour begins to fade, then decinormal hydrochloric acid until the colour is completely discharged. A reading of the acid is now taken, methyl orange is added, and the titration continued until the liquid shows an acid reaction. The amount of acid used up during the methyl-orange titration is that required for the equation:

$$NaHCO_3 + HCl = NaCl + H_2O + CO_2.$$

From this is calculated the weight of calcium carbonate originally present in the soil.

The object of employing normal HCl in the first part of the titration is to prevent unnecessary dilution, but the liquid must be kept in motion so that the acid is never in excess at any point, with consequent evolution of carbon dioxide.

Pure caustic soda should be used for the estimation, and a blank experiment should be carried out in order to correct for :—

(i.) The amount of $CO_2$ originally present in the 100 c.c. of caustic soda used in the Reiset tower.

(ii.) The $CO_2$ present in the air contained in the absorption apparatus previous to addition of the soda solution.

(iii.) Any $CO_2$ which may be present in the acid used for decomposing the carbonates.

### "Total" Mineral Constituents.

**77.** *Hydrochloric Acid Extraction.*[1]

Fifty grams of the fine earth are boiled in an open flask with 100 to 150 c.c. of concentrated hydrochloric acid for a short time, with constant shaking, in order that the acid may attain constant strength. The flask is then loosely stoppered with a funnel or a sealed-off glass bulb, and the contents digested on the water-bath for forty-eight hours. The mixture is filtered, while hot, through a Buchner funnel, washed thoroughly with hot water, and the filtrate and washings made up to 1000 c.c.

Unignited soil must be taken for extraction, since ignition occasions drastic and variable alteration of the soil—*e.g.*, no constant proportion is found between the potash extracted from ignited and unignited soil.

**78.** *Phosphoric Acid.*

Fifty c.c. of the solution are evaporated to dryness in a platinum or porcelain dish, and ignited ; the residue

[1] If, as sometimes happens, only the phosphoric acid and potash are to be estimated, it will be sufficient to use 20 grams of soil, and make up the hydrochloric acid extract to 500 c.c.

is again taken up with hydrochloric acid, filtered, again evaporated to dryness, and heated for half an hour at 105° C. The residue is now dissolved, as far as possible, in dilute nitric acid, filtered, and made up to about 50 c.c. To the solution is added several grams of ammonium nitrate, and 50 c.c. of ammonium-molybdate solution. The mixture is allowed to stand in a warm place for twenty-four hours, the yellow precipitate filtered off, and washed thoroughly with 1 per cent. nitric acid solution.

The precipitate is dissolved in dilute ammonia, and the filtered solution reprecipitated by the addition of excess of nitric acid (1 part of acid to 2 parts of water). A further 5 c.c. of ammonium molybdate are also added, and the mixture allowed to stand in a warm place for some time, as before. The precipitate thus obtained is filtered in a Gooch crucible, washed well with 1 per cent. nitric acid, dried in the oven, and ignited until of a uniform dark blue colour and of constant weight. During ignition the base of the Gooch should be protected by a cap or by placing it inside another crucible.

The residue contains 3·96 per cent. of $P_2O_5$, which corresponds to the formula $P_2O_5 . Mo_3O_8 . 21 MoO_3$.

**79.** *Potassium.*

Fifty c.c. of the original soil extract from **77** are taken, 10 c.c. of a 5 per cent. barium hydroxide solution added to precipitate any sulphate, and the mixture evaporated to dryness in a dish, and then ignited to dull redness for fifteen minutes, iron, alumina, etc., being thus rendered insoluble. The soluble alkali salts are then extracted by three or four successive treatments with boiling water, the residue being broken up with a glass rod during extraction. To the aqueous extract are now added 2·5 c.c. of perchloric acid solution of specific gravity 1·12 (20 per cent.) and the mixture evaporated on a sand-bath until there is a vigorous evolution of dense white fumes of perchloric acid.

When cool, 20 c.c. of 95 per cent. alcohol are stirred in, and after settling, the clear solution decanted off through a dried weighed filter paper. To the residue are then added 10 c.c. of 95 per cent. alcohol containing ·2 per cent. of perchloric acid, the insoluble potassium perchlorate transferred to the weighed filter paper, a further 20 or 30 c.c. of the alcohol containing ·2 per cent. perchloric acid being employed for washing the last traces on to the filter. The washing out of the perchloric acid from the filter paper and contents is effected by using a minimum of 95 per cent. alcohol, the washings being tested with litmus paper until free from acidity. The filter paper and precipitate while still in the funnel are dried in the steam-bath for twenty minutes, the paper and its contents being then placed in a tared weighing bottle, and dried at 100 to constant weight (1 part $KClO_4$ = ·3401 part $K_2O$).

80. *Iron and Aluminium.*

One hundred c.c. of the original extract, equivalent to 5 grams of soil, are heated in a beaker nearly to the boiling point, with a few c.c. of nitric acid, in order to oxidise any ferrous iron to the ferric condition. Strong ammonia solution is then added until precipitation is complete. The mixture is boiled for five minutes or so, allowed to settle, and the clear solution decanted off through a filter. The precipitate is again washed by decantation with 50 c.c. or so of boiling water, as much of the clear solution poured off as possible, and the residue dissolved in a few c.c. of hydrochloric acid, and precipitated with ammonia solution as before, excess of ammonia being avoided. The precipitate is transferred to the filter, washed free from chlorides with hot water, and then dried and ignited in the usual way. It consists of the ferric oxide, alumina, and phosphoric acid.

The iron is estimated separately, the phosphoric acid having been determined as above, and the alumina estimated by difference.

### 81. *Iron.*

One hundred c.c. of the original extract are boiled with nitric acid if any ferrous iron is present; if not, the solution is precipitated forthwith by strong ammonia, the precipitate filtered and washed, and then dissolved through the filter paper with dilute sulphuric acid. The solution is transferred to a small conical flask fitted with a singly bored stopper and delivery tube bent twice at right angles, the end dipping under distilled water in a beaker. Pure zinc is added to the liquid in quantity sufficient to reduce all the iron present, and the latter is then determined by titration with potassium permanganate solution.

A weighed quantity (about 5 grams) of zinc should be used for the reduction, and blank experiments performed in order to allow for any iron with which it may be contaminated.

### 82. *Manganese.*

The filtrate from the iron and aluminium determination (80) is concentrated to about 200 c.c. Strong ammonia is added until the liquid is alkaline, and then bromine water is run in until the colour persists after stirring. The solution is heated to boiling, the beaker being kept covered with a clock-glass; as the bromine escapes the beaker is allowed to cool somewhat, more ammonia and bromine water added, and the mixture then heated as before. This process is continued until the manganese is completely precipitated, which requires from fifteen to thirty minutes. The precipitate is filtered off while still hot, washed with boiling water, dried, ignited, and weighed as $Mn_3O_4$.

83. *Calcium.*

The filtrate from the manganese is concentrated to 100 c.c. or so, if necessary, rendered alkaline with ammonia, and while still boiling treated with saturated ammonium oxalate solution so long as any precipitate is formed. Too great excess of the precipitant is to be avoided. After precipitation the liquid is boiled for ten minutes, allowed to cool a little, filtered, and washed with hot water. The precipitate is dried in the oven and strongly ignited, first over the Bunsen flame and then by means of the blowpipe, and finally weighed as CaO.

84. *Magnesium.*

The filtrate and washings from estimation of the calcium are concentrated to about 150 c.c., allowed to cool, and transferred to a conical flask of about 800 c.c. capacity. Sufficient sodium-phosphate solution ($Na_2HPO_4$) is added to precipitate the magnesium, and then about 10 c.c. of strong ammonia. The liquid is allowed to stand in the cold for twenty-four hours; the precipitate formed is filtered off and washed with 2 per cent. ammonia solution until free from chlorides, dried, and ignited, first at moderate heat and then very strongly. The residue is magnesium pyrophosphate ($Mg_2P_2O_7$), and from its weight is calculated the percentage of MgO.

85. *Sulphuric Acid.*

For the determination of the sulphates, 200 c.c. of the original hydrochloric acid extract are evaporated to about 50 c.c., a little solid barium chloride added, and the beaker heated on the water-bath for two or three hours. After standing for another twelve hours the small precipitate is filtered off, washed thoroughly with hot water, dried, and ignited together with the filter. The residue is $BaSO_4$, and from its weight the percentage of sulphuric acid is calculated as $SO_3$.

**86. Nitrates.**

If it is desirable to estimate the nitrates in the soil, the following method should be adopted.

The scil sample must be fresh, and immediately after sampling is rapidly dried by artificial heat in order to prevent the production of further quantities of nitrates, which are otherwise readily formed during manipulation. About 500 grams of the soil are pressed into a large Buchner funnel connected with the pump; 50 c.c. of hot distilled water are poured over the sample, allowed to stand for a few minutes, and then sucked through. This operation is repeated with successive small quantities of water until all the nitrate has been washed out. This should be done with about 200 c.c. of water.

The liquid extract is filtered, the nitrates reduced to ammonia by means of the copper-zinc couple (**48** (*d*)), and the ammonia estimated after distillation, either by direct titration or with Nessler's solution (**212**), according to the amount.

### "Available" Phosphoric Acid and Potash.

As an attempt to differentiate between the "total" plant foods and that proportion of them which can be readily utilised, the "available" phosphorus and potassium are estimated by extracting the soil with dilute citric acid. The idea underlying this procedure is an attempt to reproduce in the laboratory the natural solvent agencies operative in the soil which govern the absorption of the more readily available plant nutrients.

**87.** *Citric Acid Extraction.*

Two hundred grams of the "air-dried fine earth" are placed in a Winchester quart bottle, together with 20 grams of crystallised citric acid and 2 litres of water. Old acid bottles should be used for this purpose. The mixture is shaken thoroughly from time to time during a period of seven days. If a mechanical shaker is used, twenty-four hours will be sufficient.

The soil is now filtered off, and two portions of the filtrate of 500 c.c. each are evaporated to dryness and ignited, to remove the citric acid and other organic matter. The residues are extracted with hydrochloric acid, again evaporated to dryness, and heated for some time at 105° C., in order to render the silica insoluble. In one portion is estimated the phosphoric acid, and in the other the potash, by the methods described above under the "Total Minerals."

## Statement of Results.

The results of the chemical analysis of soil should be stated as shown in the following typical example :—

**88.** *Alluvial Soil (Kent).*

| | Per cent. |
|---|---|
| Moisture . . . . . . | 3·71 |
| Loss on Ignition . . . . | 6·22 |
| Nitrogen . . . . . . | 0·334 |
| Alumina, $Al_2O_3$ . . . . . | 1·47 |
| Ferric Oxide, $Fe_2O_3$ . . . | 2·31 |
| Manganese Oxide, $Mn_3O_4$ . . | trace |
| Magnesia, MgO . . . . | 0·29 |
| Lime, CaO . . . . . | 1·30 |
| Carbonates (as $CaCO_3$) . . . | 2·53 |
| Potash, $K_2O$ . . . . . | 1·12 |
| ,, (available) . . . . | 0·043 |
| Phosphoric Acid, $P_2O_5$ . . . | 0·134 |
| ,, ,, (available) . . | 0·013 |
| Sulphuric Acid, $SO_3$ . . . . | 0·06 |

# SECTION III.—FERTILISERS AND MANURES

## CHAPTER X

### ARTIFICIAL NITROGENOUS MANURES

#### Sulphate of Ammonia.

AMMONIUM sulphate is obtained from the ammoniacal liquor obtained during the distillation of coal. The ammonia liquor is either saturated directly with sulphuric acid, or else it is distilled with lime and the gaseous ammonia collected in sulphuric acid. The latter method yields the ordinary pure sulphate of ammonia of commerce, direct saturation giving a very inferior product containing sulphides, thiocyanates, and other injurious substances.

Sulphate of ammonia should be 95 per cent purity, and normally contains 24·5 per cent. ammonia or 20·2 per cent. nitrogen.

**89.** *Ammoniacal Liquor.*

(*a*) *Qualitative Examination.*—The ammonia liquor may be examined qualitatively as follows:—

| | |
|---|---|
| Sulphides . . | $H_2S$ evolved on addition of acids. |
| Thiosulphates . | S precipitated on addition of acids. |
| Carbonates . . | $CO_2$ formed on addition of acids. |
| Sulphates . . | White precipitate with barium chloride in presence of HCl. |

| | |
|---|---|
| Chlorides . . | Test with silver nitrate in acid solution. |
| Thiocyanates . | Blood-red colour produced with ferric chloride. |
| Ferrocyanides . | Prussian blue formed with ferric chloride. |

All of these substances are likely to be present, as well as tarry matter, organic bases, etc.

(*b*) *Total Ammonia.*—Ammonia in ammoniacal liquor amounts in the average to 2 per cent. It is estimated by distilling 20 c.c. of the liquor with caustic soda, and collecting the ammonia evolved in 50 c.c. of $\frac{N}{1}$ $H_2SO_4$. The apparatus used is that previously described in the Kjeldahl estimation (7, p. 12).

The free ammonia may be estimated by similarly distilling 20 c.c. of the liquor without the addition of caustic soda.

**90.** *Sulphate of Ammonia.*

A commercial sample of the fertiliser is chosen, its colour and general appearance noted, and a small portion heated on platinum foil in the Bunsen flame. A good sample should volatilise without leaving any residue.

(*a*) *Sand and Insoluble Material.*—Ten grams of the fertiliser are weighed out, dissolved in water in a beaker, and filtered into a 250 c.c. flask. The residue, if any, is washed thoroughly, and the filtrate and washings made up to the mark. The insoluble material is dried in the oven, and ignited in a crucible together with the filter paper, and its amount expressed in percentage of the original material.

(*b*) *Ammonia.*—Twenty-five c.c. of the solution from (*a*) are distilled in the Kjeldahl distillation apparatus with caustic soda, and the volatilised ammonia collected in 25 c.c. $\frac{N}{2}$ sulphuric acid and estimated by back-titration in the usual manner.

As an *alternative method*, dispensing with distillation, 25 c.c. of the solution may be boiled in a conical flask with 50 c.c. of $\frac{N}{2}$ sodium hydroxide. A small funnel is placed in the neck of the flask to prevent loss by spirting, and boiling is continued until no more ammonia is given off. This may be determined by holding a strip of red litmus paper in the issuing steam. When the reaction is complete the liquid is cooled down and the funnel and neck of the flask are rinsed back with distilled water. A drop or two of methyl orange is added, and the residual caustic soda is determined by running in $\frac{N}{2}$ $H_2SO_4$ until neutral.

$$(NH_4)_2SO_4 + 2NaOH = Na_2SO_4 + 2H_2O + 2NH_3.$$

Every 40 grams of NaOH used up during the reaction corresponds to 17 grams of ammonia or 14 grams of nitrogen.

(*c*) *Qualitative Examination.*—The sulphate of ammonia solution should meanwhile be examined for the constituents noted above under the Ammoniacal liquor (**89** (*a*), p. 105). If any of these is present in large quantity, which is very unlikely, it should be estimated separately.

## Nitrate of Soda.

Nitrate of soda is obtained from extensive natural deposits occurring in South America. The crude material is known as *caliche*, and may contain 20 to 60 per cent. of sodium nitrate. It is purified by recrystallisation, and the finished product contains in the average 96 per cent. sodium nitrate.

Commercial nitrate of soda may contain up to 1 per cent. of sodium chloride, small quantities of sulphates, and occasionally a little sodium iodate. The fertiliser is very hygroscopic, and the estimation of moisture is thus of some importance.

**91.** *Qualitative Examination of Nitrate of Soda.*

The sample should be examined qualitatively for *sulphates* and *chlorides* in the usual manner, and observation made of any insoluble or difficultly soluble matter present.

*Iodates* should be tested for as follows:—About 0·5 gram of the nitrate of soda is dissolved in a little water, acidified with dilute hydrochloric acid, and then mixed with a solution of sulphurous acid. The iodic acid is thus reduced with the formation of free iodine, which will give a blue coloration on the addition of starch paste.

$$2NaIO_3 + 2HCl + 5H_2SO_3 = 2NaCl + 5H_2SO_4 + I_2 + H_2O.$$

**92.** *Analysis of Nitrate of Soda.*

(*a*) *Moisture.*—Two or 3 grams of the well-drawn fair sample are weighed into a porcelain dish and dried to constant weight in the steam oven. The loss in weight is moisture.

(*b*) *Insoluble Matter and Sand.*—Twenty grams of the manure are dissolved in water, the insoluble matter collected on a filter paper and well washed. The residue is dried and ignited together with the filter paper in a weighed crucible, and is weighed as sand.

The filtrate and washings from the insoluble material are made up to 1 litre with distilled water and the solution retained for the following estimations.

(*c*) *Chlorides.* — The chlorides present are best estimated volumetrically by standard silver nitrate solution, using potassium chromate as indicator. One hundred or 200 c.c. of the solution from (*b*) are taken for analysis according to the amount of chlorides present. A small quantity of potassium chromate solution is added, and the flask placed on a white tile. Decinormal silver nitrate solution (1 c.c. = 0·00585 gram NaCl) is now run in from a burette, and the liquid well shaken to ensure thorough mixing. The titration is complete when the faint reddish tinge, due to the formation of silver chromate, is permanent.

The end-point is more readily observed by using for comparison a control vessel containing the various materials with the silver not present in excess. The slightest formation of a red tint may then be easily seen, by comparison of the two solutions.

*Example:*—Vol. of solution used = 200 c.c. (*i.e.* 4 grams of fertiliser).

Vol. of $AgNO_3$ required = 5·8 c.c.

∴ Weight of NaCl present in 100 c.c. = 5·8 × 0·00585 gram = 0·0339 gram.

∴ *Percentage of NaCl in the Nitrate* = 0·85

(*d*) *Sulphates.*—If the qualitative examination shows the presence of large quantities of sulphate, 100 c.c. of the solution from (*b*) are acidified with hydrochloric acid, mixed with some ammonium chloride solution, and brought to the boil. Hot barium chloride solution is run in until all the sulphate is precipitated, the barium sulphate is filtered off, washed thoroughly, dried, and ignited. From its weight is calculated the percentage of $Na_2SO_4$ in the nitrate of soda. One part of $BaSO_4$ is equivalent to 0·609 part of $Na_2SO_4$.

(*e*) *Nitrogen.*—The quantity of sodium nitrate in the manure is generally determined by difference from 100 by subtracting the percentages of moisture, sand, sodium chloride, and sodium sulphate found. If the result is to be returned in terms of nitrogen, 1 part of $NaNO_3$ contains 0·164 part of N.

If it is desirable actually to estimate the nitrogen present, this should be done by one of the methods described below.

*Estimation of Nitrogen in Sodium or Potassium Nitrate.*

**93.** (*a*) *Nitrometer Method*—When strong sulphuric acid is added to the solution of a nitrate in presence of mercury, the nitrogen is liberated as nitric oxide.

$$2KNO_3 + 3Hg_2SO_4 + 4H_2SO_4 = K_2SO_4 + 6HgSO_4 + 2NO + 4H_2O.$$

The percentage of nitrogen is calculated from the amount of liberated NO. One litre of NO at N.T.P.[1] weighs 1·344 grams.

*Apparatus.*—The Lunge nitrometer (Fig. 18) consists of two vertical glass tubes, A and B, connected at the foot by a length of stout-walled rubber tubing, and attached by clamps to a suitable stand. The tube B is plain, and is used as a levelling tube, but A is graduated from the top downwards, usually to a total volume of 50 c.c., reading to $\frac{1}{10}$ c.c. The top of this tube is contracted and ends in a three-way tap, by means of which it can be connected either with the reservoir *d* or the capillary tube *e*. It may also be shut off from both.

*Procedure.* — The tube B is raised and mercury poured in until it just reaches the glass tap in A and fills the bottom 2 inches of B. This condition is obtained by raising or lowering B as required. The tap is now closed and the levelling tube clamped in position.

About 0·1 to 0·2 gram of the sample of nitrate of

[1] Normal temperature and pressure, 0° C. and 760 mm.

soda is weighed out accurately and washed into the funnel *d* by means of 2 to 4 c.c. of water. When the material is dissolved the funnel is connected to the tube A, and by cautiously lowering B the liquid is drawn into the nitrometer. The funnel is washed out with 1 c.c. or so of water from a wash-bottle with a fine jet, and the washings also drawn into A. The sides of the funnel are now rinsed with 3 to 4 c.c. of concentrated $H_2SO_4$, which is then passed into the tube by lowering B. This is quickly repeated twice with small quantities of sulphuric acid, the tap being closed immediately the washings have been transferred. If any air is enclosed during these operations it must be immediately removed by raising B slightly, gently opening the tap and closing it again when the air has escaped.

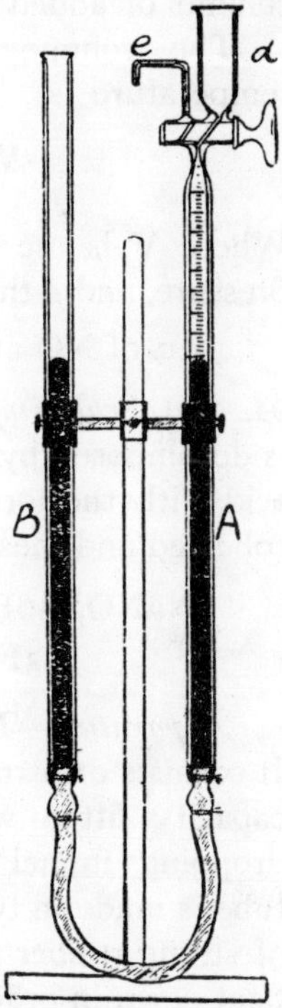

FIG. 18.—The Lunge nitrometer.

The graduated tube is now unclamped, the connecting rubber pinched between the fingers close to the lower end of A, and the latter inclined in both hands almost horizontally and with a quick motion brought back to the vertical. The process is repeated for two or three minutes so that the mercury and the acid are well mixed. The tubes are then adjusted in the clamps so that the mercury is approximately level in both limbs, and the apparatus allowed to stand till the temperature coincides with that of the room.

To read off the volume of gas, the tube B is carefully adjusted so that the mercury in it is higher than that in A by 1 mm. for every 6·5 mm. of solution in A. This makes the necessary allowance for the short column of added liquid.

The volume of gas corrected for pressure and temperature is

$$V \times \frac{(P-p) \times 273}{760 \times (t+273)} \text{ c.c.}$$

Where V is the observed volume, P the barometric pressure, and $p$ the tension of aqueous vapour at $t°$ C.

1 c.c. of NO at N.T.P. = 0·000627 gram nitrogen.

**94.** (*b*) *Schlössing's Method.*—In this process the nitrate is decomposed by ferrous sulphate and strong sulphuric acid with the formation of nitric oxide. The latter is collected and measured.

$$2NaNO_3 + 6FeSO_4 + 5H_2SO_4$$
$$= 3Fe_2(SO_4)_3 + 2NaHSO_4 + 4H_2O + 2NO.$$

*Apparatus.*—The apparatus is shown in Fig. 19. It consists of a round-bottomed flask of about 200 c.c. capacity fitted with a two-holed stopper carrying a dropping funnel and a delivery tube. The delivery tube is made in two parts, joined together with a piece of strong rubber tubing fitted with a screw-clip, and is bent down in order to dip under the surface of water in a trough in which is inverted a measuring tube full of water.

*Procedure.*—Twenty c.c. of a 10 per cent. solution of ferrous sulphate are placed in the flask and mixed with 50 c.c. of strong $H_2SO_4$. The apparatus is fitted together, the clip opened, and the solution boiled until all the air has been expelled. The screw-clip is now closed and at the same time the flame removed. If the apparatus is tight, water will rush up and fill the delivery tube as far as the tap.

The measuring tube is now fixed in position over the end of the delivery tube and about 0·2 gram of sodium nitrate (or 10 c.c. of the solution from **92** (*b*), p. 108) dissolved in water and placed in the dropping funnel. By carefully opening the funnel the nitrate is drawn into the flask and the last traces similarly washed in with a few c.c. of water, care being taken to prevent the introduction of any air. The mixture is now heated

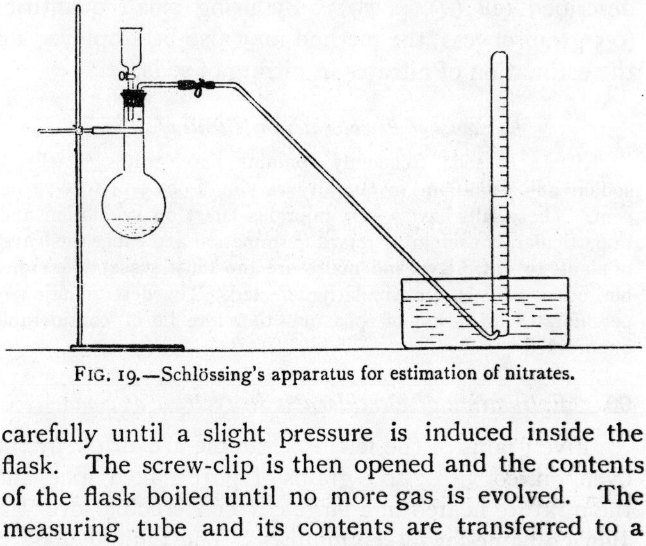

FIG. 19.—Schlössing's apparatus for estimation of nitrates.

carefully until a slight pressure is induced inside the flask. The screw-clip is then opened and the contents of the flask boiled until no more gas is evolved. The measuring tube and its contents are transferred to a cylinder of water and the volume of the nitric oxide which has been collected read off after bringing the water to the same level inside and outside the tube. The volume of the gas is corrected for temperature and pressure and the calculation of the percentage of nitrogen made as in the nitrometer method (*a*).

The collection of the gas should really be made over

mercury instead of water. To allow for the errors introduced by the use of water it is advisable to control the result by carrying out an experiment, under exactly the same conditions, with chemically pure sodium or potassium nitrate. The purity of the fertiliser under examination may then be measured by direct comparison without further calculation.

**95.** (*c*) *Estimation by Reduction to Ammonia.*—With manures containing only small quantities of nitrates it is advisable to use the reduction process already described (**48** (*d*), p. 70). By using small quantities (0·5 gram or less) the method may also be employed for the estimation of nitrates in nitrate of soda, etc.

### *Existence of Perchlorates in Nitrate of Soda.*

Nitrate of soda frequently contains perchlorates (chiefly of sodium and potassium) in quantity varying from 0·14 to 6·79 per cent. These salts have a very injurious effect on vegetation and, in particular, considerably retard germination and cause the leaves of plants to wilt. Rye and maize are the most susceptible crops, but other cereals are similarly affected. The determination of perchlorates in nitrate of soda may therefore be of considerable importance.

### **96.** *Estimation of Perchlorates in Nitrate of Soda.*

Five grams of the fertiliser sample are dried in the oven, mixed with 7 to 8 grams of pure slaked lime, and the mixture heated in a large covered crucible over the Bunsen flame for fifteen minutes. The ignited mass is extracted with water, filtered, and thoroughly washed, and the filtrate then neutralised exactly with nitric acid. The total chlorine present, which is made up of that originally present as chloride in the fertiliser, together with that derived from the perchlorate, is then estimated in the solution by direct precipitation, or volumetrically by titration with silver nitrate, as in para. **92** (*c*).

## Nitrate of Lime.

Nitrate of lime is prepared from atmospheric nitrogen by a modern extension of the original Cavendish process. The product is generally clean and free from extraneous matter, and averages 13 per cent. nitrogen. It can be valued on exactly the same basis as nitrate of soda, and its analysis is carried out in a similar manner. Its chief disadvantage is its hygroscopicity; it absorbs moisture from the atmosphere more readily than any other manure in common use, and is therefore generally sold in hermetically sealed drums.

**97.** *Hygroscopicity of Nitrate of Lime.*

(*a*) A quantity of the fertiliser on exposure to a moist atmosphere quickly becomes moist, then pasty, and finally liquid.

(*b*) *Comparative Hygroscopicity.*—Into three clock-glasses are weighed equal quantities of nitrate of lime, nitrate of soda, and sulphate of ammonia. The clock-glasses are placed under a large bell-jar, in which is placed a beaker of water to keep the enclosed atmosphere of a constant moisture content. The dishes should be weighed every two or three hours, and the increases in weight, showing the amounts of water absorbed, plotted out graphically for comparison.

## Calcium Cyanamide or Nitrolim.

This material is prepared from atmospheric nitrogen by combination with heated calcium carbide. Chemically it consists essentially of $CaCN_2$, but it also contains excess of lime, about 20 per cent. free carbon, and small quantities of calcium carbide—$CaC_2$. In the presence of water, as for example in the soil, the cyanamide gradually decomposes into free ammonia and calcium carbonate[1]:

$$CaCN_2 + 3H_2O = 2NH_3 + CaCO_3.$$

It consequently acts as a basic manure. Commercial nitrolim contains about 20 per cent. of nitrogen, which is in a fairly readily available condition. Its chief drawbacks are its light powdery condition and its frequent slight content of calcium carbide and

[1] For decomposition of calcium cyanamide in the soil, see *J. Agric. Sci.*, vol. ix., p. 113.

phosphide, which produce noxious gases on moistening with water. For this latter reason it should be incorporated with the soil a week or so before any seed is sown.

**98.** *Qualitative Examination.*

(*a*) Some nitrolim is moistened with water in an open dish. The smell of acetylene generally produced may be noted, and the gases formed should be tested for free ammonia by means of litmus and turmeric paper. During the moistening process the manure will be found to heat up, owing to the slaking of the free lime it contains.

(*b*) A small portion of the manure is fused with three times its bulk of common salt. The mass is extracted with water and filtered. The presence of *cyanide* in the solution may be readily shown by adding a few drops each of ferrous sulphate and ferric chloride. On acidification with HCl a precipitate of prussian blue is formed.

**99.** *Analysis of Nitrolim.*

(*a*) *Nitrogen.*—The nitrogen is best estimated by digestion with strong sulphuric acid by the ordinary Kjeldahl process.

(*b*) *Carbon and Insoluble Matter.*—About 2 grams of the manure are warmed with 50 to 100 c.c. of water, and the mixture filtered on a tared filter paper while still hot. The undissolved material is washed back into the beaker and boiled with dilute hydrochloric acid for a few minutes, and the residue again transferred to the filter paper, where it is washed with dilute acid until the filtrate no longer contains calcium salts in solution. The residue is then washed well with water, dried in the steam oven, and weighed. This gives the carbon and insoluble matter together. Filter paper and contents are now moistened with ammonium nitrate solution and ignited until the whole of the carbon is removed. The weight of the residual insoluble material gives the amount of carbon by difference from the first weighing.

Commercial cyanamide contains on the average about 14 per cent. of free carbon, and about 6 per cent. of insoluble material—chiefly alumina, iron oxide, and silicious matter.

# CHAPTER XI

## Organic Nitrogenous Manures

### Farmyard Manure.

Farmyard and stable manures are mixtures of urine, excrement, waste portions of fodder, and the litter used for the animals. All of these bodies contain nitrogen, phosphates, and potash; but an ordinary analysis is a difficult proceeding, owing to the great patience and care which must be exercised in sampling. Besides this, one of the most valuable assets of farmyard manure is its mechanical condition, which is difficult to gauge; it introduces into the soil large quantities of humus bodies, whereby the physical state of the soil is profoundly modified, generally for its improvement.

**100.** *Examination of Farmyard Manure.*

(*a*) *Nitrogen.*—The nitrogen in farmyard manure exists in all stages of complexity, from undecomposed protein material down to ammonium compounds.

*Ammonium Salts, Amides, etc.*—Fresh manure frequently smells distinctly of ammonia and reacts alkaline to litmus paper; older manure is more nearly neutral in character. On heating with dilute (5 per cent.) caustic soda, free ammonia is liberated from the ammonium salts and the amides present, and the ammonia may be recognised in the usual way.

*Protein Material.*—The material remaining after heating the dung with dilute soda and driving off all the loosely combined nitrogen, is dried and heated in a tube with soda-lime, when more ammonia will be produced, showing the presence of complex nitrogenous bodies.

(*b*) *Phosphoric Acid and Potash.*—Twenty to 30 grams of the manure are extracted by boiling with hydrochloric acid. The solution is filtered, and the potassium salts and phosphates may be recognised in the filtrate by the usual tests.

(*c*) *Humus.*—If this has not already been done (p. 60), humus should be prepared from well-rotted manure by the method already described (*loc. cit.*).

### Liquid Excrement.

The urine of animals, which is largely absorbed by their litter and is thus a constituent of fresh farmyard manure, is one of the latter's most valuable constituents. It contains simple nitrogen compounds like urea, uric acid, and ammonia, as well as soluble phosphates and potassium salts, the various constituents being almost ready for plant nutrition and requiring but slight further changes actually to become so. The concentration of the liquid excrement of the various animals in manurial constituents increases in the order—pig, sheep, cow, horse.

**101.** *Separation of Urea from Human Urine.*

Urea, $CO(NH_2)_2$, the most important constituent of urine, is prepared from it as follows:—About 2 litres of urine, containing about 30 to 40 grams of urea, are evaporated down to small bulk (200 or 300 c.c.), allowed to cool, and mixed with 50 c.c. of nitric acid. If no precipitation occurs the liquid is further evaporated until urea nitrate separates on cooling. The precipitated nitrate is filtered off, dissolved in water, and 1 gram of potassium permanganate added in order to destroy the colouring matter. Barium carbonate is now added to neutralise the excess of acid and decompose the urea nitrate:

$$2(CON_2H_4.HNO_3)+BaCO_3$$
$$=2CON_2H_4+Ba(NO_3)_2+H_2O+CO_2.$$

The mixture is now evaporated to complete dryness on the water-bath and the residue extracted with alcohol, which removes only the urea. On concentrating the alcoholic solution the urea separates in the form of rhombic prisms, which readily dissolve in water.

**102.** *Experiments with Urea.*

(*a*) A small quantity of urea is strongly heated in a porcelain or platinum dish, whereby decomposition takes place, ammonia being liberated and biuret formed:

$$2CO(NH_2)_2=\underset{\text{Biuret.}}{NH_2.CO.NH.CO.NH_2}+NH_3.$$

The residue is dissolved in dilute caustic soda and treated with a few drops of copper-sulphate solution; a beautiful violet-red coloration is produced. This is the "biuret reaction" frequently used as a characteristic test for the proteins (see **9**, *c*, p. 15).

(*b*) A small quantity of urea is heated with caustic-soda solution. It decomposes with the formation of ammonia and carbon dioxide:

$$CO(NH_2)_2+H_2O=CO_2+2NH_3.$$

A similar reaction proceeds in the manure heap, on exposing urine to the air and in the soil, under the influence of bacteria of the type of *Micrococcus ureæ*, ammonium carbonate being formed intermediately:

$$CO\begin{cases}NH_2\\NH_2\end{cases} + 2H_2O = CO_3\begin{cases}NH_4\\NH_4\end{cases} = 2NH_3 + CO_2 + H_2O$$

Urea. Ammonium Carbonate.

**103.** *Uric Acid.*

Uric acid is normally present in small quantity in the urine of carnivora and of man, and separates from urine on allowing to stand for some time. It may be detected by the *murexide* reaction: a small quantity of uric acid is evaporated to dryness with a few drops of nitric acid. On moistening the residue with ammonia, a purple colour is produced, due to the formation of murexide, a purple dye which was formerly used as such.

**104.** *Hippuric Acid.*

This material takes the place of uric acid in the urine of herbivora, and particularly in that of the horse, in which it exists to the extent of 2 per cent. It may be tested for by evaporating the urine nearly to dryness and boiling for some time with caustic soda until no more ammonia is liberated. The solution will now contain benzoic acid, formed by hydrolysis of the hippuric acid. The solution is acidified with sulphuric acid, boiled with animal charcoal, and filtered hot.

On testing the concentrated filtrate with ferric chloride a buff-coloured precipitate of ferric benzoate should be obtained.

## Shoddies, Wool Wastes, etc.

Numerous waste products containing nitrogenous material are largely used as manures, particularly for fruit and hops. These waste products are frequently derived from the textile industries, and comprise wool, silk, hair, and fur wastes. They vary greatly in quality, and a true shoddy consisting of waste short-staple wool may contain 14 per cent. or over of nitrogen, whereas wool and flock dusts, factory sweepings, etc., may be so mixed with dirt as only to contain 3 or 4 per cent. of nitrogen. The "shoddies" also vary considerably in value.

Besides these materials there are also cloth and rag wastes, skin waste, feathers, hoof and horn shavings, rabbit flick, slaughter-house refuse, leather cuttings and dust, and many others. This latter class is generally much less valuable than the fibre wastes, being less readily decomposed in the soil and usually much coarser in texture.

### **105.** *Qualitative Examination of Fibrous Waste Products—Shoddies, etc.*

(*a*) *Mechanical condition.* — The texture of the material should be carefully noted, and any obviously extraneous matter picked out and examined separately.

(*b*) *Sandy matter.*—If no analysis is to be carried out (see below) the material should be well fingered, to detect sand, and a few grams ignited and the residue examined; it should be grey or practically colourless in appearance if purely animal or vegetable in character, and as a rule only if mixed with mineral substances will it contain much iron and have a reddish colour.

The presence of much oily matter may be deleterious and prevent the bacterial decomposition of the manure. Large quantities may be detected as follows:—

(*c*) *Oil.*—A few grams of the manure are twice shaken up with ether in a tube, the ethereal solutions united and filtered, and then allowed spontaneously to evaporate. The oily residue, if any, may be examined by the tests described in **13, 14**, pp. 22, 24.

(*d*) *Nature of the Fibre.*—Wool and silk, consisting of almost pure protein (17 per cent. N), are the valuable constituents of textile waste products. Frequently, however, they are mixed with quantities of cellular material like cotton and linen fibre which has no fertilising action. Small portions of the sample, or suspicious-looking fibres taken from it, should be subjected to the following tests, which should also be carried out with pure material :—

| Tests. | Wool. | Silk. | Cellulose (Cotton or Linen). |
|---|---|---|---|
| Burn in Bunsen flame. | Do not burn readily ; char, sinter, and frizzle up, giving characteristic smell like burnt hair. | | Burns readily with clear flame. Smells somewhat like burning brown paper. |
| Digest with Schweitzer's reagent—(ammoniacal cupric hydroxide solution). | Insoluble. | | Soluble. |
| Boil with dilute caustic soda solution (10 per cent.). | Soluble. | | Insoluble. |
| If the fibre or fabric is light coloured, soak in a solution of magenta, or other aniline dye,* and then rinse thoroughly with warm water. | Dyed permanently. | | Does not retain the colouring matter. |
| Digest with a solution of basic lead acetate, to which has been added just enough NaOH to redissolve the precipitate first formed. | Turns brown. | Unchanged. | Unchanged. |
| Soak for some time in cold concentrated hydrochloric acid. | Insoluble. | Soluble. | Insoluble. |

* Except direct cotton dyes like those derived from benzidene.

The above tests may be carried out on a mixed fabric, if obtainable. By soaking in cold strong HCl for some time the silk is removed. After washing, the remainder is boiled with 10 per cent. NaOH, which dissolves out the woollen fibres. The remainder is cotton or linen.

**106.** *Analysis of Fibrous Waste Products.*

(*a*) *Sampling.*—As large a parcel as possible should be well mixed, spread out, and fairly sampled for analysis. Coarse portions of fabric, etc., must be cut into small pieces and distributed throughout the mass.

(*b*) *Moisture.*—Measured by drying at 100° C.

(*c*) *Sandy Material.*—Obtained, together with the ash, by incinerating a few grams of material and strongly igniting until the residue is free from carbonaceous matter.

(*d*) *Oil.*—About 10 grams of the material are thoroughly extracted with light petroleum in a Soxhlet apparatus into a weighed flask. The solvent is removed and the residue dried and weighed (see **12**, p. 19).

(*e*) *Nitrogen.*—Estimated by the Kjeldahl process.

**107.** *Examination of other Nitrogenous Waste Products.*

(*a*) *Sampling.*—The difficulty of examining products like skin and fur wastes, rabbit flick, slaughter-house refuse, etc., generally consists in the sampling. Very large pieces must be removed and the proportion measured, since they will have little manurial value. Smaller pieces must be cut up and incorporated with the main bulk previous to sampling.

Leather dust, horn and hoof dust or shavings, dried blood, etc., are generally easy to sample.

(*b*) *Qualitative Examination.*—As many organic waste products as are obtainable should be examined, and their general appearance, odour, and other characteristics noted. Particular attention should be paid to the texture, since coarse material is not nearly so valuable as that which is disintegrated. A few of the manures should be subjected to the same qualitative tests as are described for the shoddies, etc. (**105**, p. 121).

(*c*) *Analysis.*—The analysis follows the same lines as those given for the fibrous wastes, but a larger portion should be used for the nitrogen estimation (about 10 grams), the sulphuric acid digest made up to a definite volume, and aliquot parts taken for the distillation with caustic soda. This reduces the error due to the difficulty of properly sampling very mixed or coarse waste products.

# CHAPTER XII

## PHOSPHATIC MANURES

WITH the exception of the phosphates of sodium and potassium, which are occasionally used by market gardeners, etc., all the phosphatic manures in common use are calcium compounds of phosphoric acid, derived either from animal or mineral sources. Phosphoric acid is tribasic, and forms with calcium the compounds:—

| | |
|---|---|
| $Ca_3(PO_4)_2$ | Tricalcium or tricalcic phosphate |
| $Ca_2H_2(PO_4)_2$ or $CaHPO_4$ | Dicalcium hydrogen phosphate |
| $CaH_4(PO_4)_2$ | Monocalcium phosphate |

Basic phosphates are also known.

The above-named phosphates occur in the different phosphatic fertilisers, sometimes singly, sometimes mixed with one or both of the others. They are distinguished by their solubilities, as follows:—

| | Formula. | Solubility in | | |
|---|---|---|---|---|
| | | Water. | Dilute Citric acid. | Hydrochloric acid. |
| Tricalcium phosphate, "insoluble phosphate" . . | $Ca_3(PO_4)_2$ | Insoluble | Insoluble | Soluble |
| Dicalcium phosphate, "precipitated" or "reverted" phosphate . . . | $CaHPO_4$ | Insoluble | Soluble | Soluble |
| Monocalcium phosphate, "soluble phosphate" . . | $CaH_4(PO_4)_2$ | Soluble | Soluble | Soluble |

**108.** *The Calcium Phosphates.*

(*a*) *Tricalcium Phosphate.*—The amorphous salt is precipitated by adding sodium phosphate solution ($Na_2HPO_4$) in excess to a solution of calcium chloride which has been made distinctly alkaline with ammonia. The $Ca_3(PO_4)_2$ is obtained as an earthy white powder by filtering off the gelatinous precipitate, washing thoroughly, and drying in the oven.

(*b*) *Dicalcium Phosphate.*—This salt, $Ca_2H_2(PO_4)_2$, is precipitated by the addition of a solution of calcium chloride to one of ordinary sodium phosphate. The precipitate is filtered off, washed, and dried.

(*c*) *Monocalcium Phosphate.*—$CaH_4(PO_4)_2$ may be obtained by dissolving the dicalcium phosphate (*b*) in phosphoric acid solution and allowing the liquid spontaneously to evaporate. The salt forms rhombic tables, which are very readily soluble in water.

**109.** *Solubilities of the Calcium Phosphates.*

(*a*) The tricalcium and dicalcium phosphates should be tested with regard to their solubilities in water, 2 per cent. citric acid solution and 5 per cent. ammonium citrate solution. A good idea of the solubility may be obtained by comparing the densities of the phospho-molybdate precipitates obtained in each case by applying the usual test for phosphates to the solutions.

(*b*) About 1 gram each of powdered bone ash, ignited tricalcium phosphate prepared from **108** (*a*), and the precipitated dicalcium phosphate (**108** (*b*)) are mixed with 100 c.c. of water, and through each suspension is passed carbon dioxide for about ten minutes. After standing for a further fifteen minutes the liquids are filtered and the various filtrates tested with ammonium-molybdate solution. From the appearance of the yellow precipitates may be gauged the amount of phosphate dissolved in each case.

### Bones and Bone Manures.

The chief constituents of bones are tricalcium phosphate, calcium carbonate, and organic matter. The latter, which may be present to the extent of 30 per cent., is composed chiefly of nitrogenous material and fat. Nitrogen may be present to the extent of 3 to 4 per cent., and exists chiefly as bone collagen or ossein The following are the most important bone compounds used as manures :—

| Name. | Preparation. | Composition. |
|---|---|---|
| 1. "Crushed bones," "half-inch bones," etc. | Fat removed by boiling or steaming. | 4 to 5 per cent. N. 43 to 50 per cent. $Ca_3(PO_4)_2$. |
| 2. Bone meal. | Ditto, but finely crushed. | ,, ,, |
| 3. Steamed bone flour. | Steamed at 50 to 60 lb. pressure. Most of the collagen removed. | 1 to 1·5 per cent. N. 50 to 60 per cent. $Ca_3(PO_4)_2$. |
| 4. Dissolved bones. | Bone meal treated with sulphuric acid. | 2 to 3 per cent. N. 12 to 18 per cent soluble phosphate. 20 to 25 per cent. insoluble phosphate. |
| 5. Bone compound. | Steamed bone flour treated with sulphuric acid, and sometimes mixed with nitrogenous material. | About 1 per cent. N. 15 to 20 per cent. soluble phosphate. 12 per cent. insoluble phosphate. |

**110.** *Examination of Bones and Bone Compounds.*

(*a*) *Bones.*—Two whole bones are weighed separately. One is then placed in the muffle and ignited for an hour at bright red heat; when cold the residue is weighed. This gives the proportion of water and organic matter present. The residue is *bone ash*, and will be required for (*d*). The second bone is immersed for a day or two in dilute hydrochloric acid. The mineral matter is dissolved out and leaves behind a soft flexible piece of *ossein.* A portion of the solution is evaporated with nitric acid in order to destroy all organic matter, the residue taken up with dilute nitric acid and tested for *phosphoric acid* with ammonium molybdate.

The ossein obtained from the second bone is washed free from acid, dried in the oven, and weighed. To a small portion is applied the xanthoproteic reaction (see **9**, *b*, p. 15), and another portion is heated with soda-lime in order to show the presence of nitrogen. The main bulk of the ossein is now boiled with distilled water for some minutes, and then left in a beaker covered with a clock-glass on the water-bath for a few hours. On allowing to cool, the solution will set to a jelly, owing to the formation of gelatine produced from the bone collagen by the action of the water. The gelatine will give a precipitate with tannic acid solution.

(*b*) *Bone Meal and Steamed Bone Flour.*—About 5 grams each of bone meal and steamed bone flour are treated with dilute hydrochloric acid and the amount of residual organic matter in each case noted. The fineness of grinding of the samples should also be observed, as on this largely depends the availability of the manures.

(*c*) *Vitriolised Bones.*—Sixty grams of bone meal

are weighed into a porcelain basin, mixed with 30 c.c. of water, and then 10 c.c. of concentrated sulphuric acid gradually added, with constant stirring. If the mixing has been judiciously carried out, the heat produced evaporates any excess of water, and the residual material will form a solid but slightly damp mass of "vitriolised" or "dissolved" bones. The material should be extracted with water, and the presence of soluble phosphates in the solution shown by testing with ammonium molybdate.

(*d*) *Bone Superphosphate.*—The mass of bone ash from (*a*) (see above) is powdered in a mortar. Sixty grams of the material are treated with 30 c.c. of water, as before, and then with 15 c.c. of strong sulphuric acid. Effervescence, due to the presence of carbonate, will take place, and a dry, friable mass of bone superphosphate will be left behind. The presence of soluble phosphate in this material is shown by boiling with a little distilled water, filtering, and testing the filtrate with molybdate solution in the usual manner.

(*e*) *Precipitated Phosphate.*—A solution is prepared by treating raw bone or bone meal with hydrochloric acid as in (*a*). To the filtered solution is added powdered chalk until no further effervescence takes place. The precipitated material is washed once or twice by decantation and then dried on the water-bath, and forms "precipitated phosphate," which is sometimes used as a manure or an adjunct to animal feeding. Its solubility should be tested with ammonium citrate solution and with distilled water.

### Rock Phosphates and Superphosphate.

The chief mineral phosphates are *apatite*, $2Ca_3(PO_4)_2 . CaF_2$, the *coprolites*, and the extensive *rock-phosphate* deposits such as are met with in Belgium, Florida, Algiers, Aruba, and elsewhere. The rock phosphates are largely used as a source of superphosphate of lime, the well-known fertiliser.

111. *Examination of Rock Phosphates.*

(*a*) Samples of finely ground rock phosphates, such as Florida or Aruba phosphates, are qualitatively examined for calcium, aluminium, and iron. A separate quantity is moistened with a little water and treated with strong sulphuric acid. The gas formed should be tested for hydrofluoric acid and also for carbon dioxide, hydrochloric acid, and $H_2S$. The residual material on cooling should set dry and friable, and may be tested for water soluble phosphoric acid.

(*b*) The solubility of the ground samples should be tested, as previously described, in distilled water, 2 per cent. citric acid solution, and ammonium citrate solution, and also by passing carbon dioxide into a suspension of the material in water.

112. *Qualitative Examination of Superphosphate.*

(*a*) A sample of superphosphate should be tested for aluminium and iron, for soluble calcium, and for soluble phosphoric acid. Its acid character will be made apparent by testing with blue litmus paper.

(*b*) A small quantity of superphosphate is mixed with some nitrate of soda or nitrate of lime, and the mixture incorporated by rubbing in a mortar. A portion of the mass is heated in a test-tube with a drop or two of water, and the remainder is allowed to stand for a day or so. Free nitric acid will be formed in both cases. This emphasises the fact that nitrates and superphosphate are incompatible manures.

## Analysis of Superphosphate of Lime.

Superphosphate, being prepared chiefly from ground mineral phosphates by the limited action of sulphuric acid, consists chiefly of $CaH_4(PO_4)_2$ and $Ca_3(PO_4)_2$, together with calcium sulphate. It also sometimes contains "reverted" or "retrograde" phosphate,

$Ca_2H_2(PO_4)_2$, and is mixed with iron and aluminium phosphates and silica, and occasionally with a small quantity of organic matter. Although a complete analysis is described below, it is more usual only to estimate the moisture and the "total" and "water soluble" phosphates.

**113.** *Moisture and Combined Water.*

Two grams of the manure are weighed on a watch-glass and heated for five hours at 100° C. in the water-oven. The loss in weight is hygroscopic moisture. The chemically combined water is determined by heating the dry manure to constant weight in the air-oven at 150° C.

**114.** *Organic Matter.*

About 3 grams of the superphosphate are weighed into a platinum dish and milk of lime added until distinctly alkaline. The mass is mixed well with a glass rod, evaporated down on the water-bath, and dried in the air-oven at 150° C. to constant weight. The dish and its contents are then ignited and again weighed. The loss represents organic matter.

The nitrogenous material (if any) may be estimated in 2 to 3 grams of the original manure by the ordinary Kjeldahl method.

**115.** *Sand, Silica, etc.*

The residue from the determination of the moisture is digested with hot dilute hydrochloric acid and filtered. The insoluble residue is washed with water, dried, and ignited. The residue is sand, etc.

**116.** *Phosphates.*

(*a*) *Water Soluble Phosphate.*—The superphosphate is ground in a mortar, and 2 grams of the powdered material are weighed out, transferred to a small mortar and rubbed up with a little water (about 30 c.c.). The clear liquid is filtered off and the residue twice subjected to the same operation, and finally boiled with a small quantity of water, the whole being then transferred to the filter, where the insoluble matter is washed until the washings are free from acid. Filtrate and washings are then made up to 250 c.c.

Fifty c.c. of the solution are pipetted into a beaker, mixed with 10 c.c. of concentrated nitric acid and then with 60 c.c. of ammonium molybdate solution. The mixture is allowed to stand in a warm place overnight, and the yellow precipitate of ammonium phospho-molybdate then filtered off and washed with 1 per cent. nitric acid solution. The precipitate is dissolved off the filter with strong ammonia, the solution nearly neutralised with hydrochloric acid and mixed with 25 c.c. of magnesia mixture.[1] The liquid is allowed to stand for twelve hours at the ordinary temperature. If all the phosphoric acid has been precipitated (determined by testing the supernatant liquid with ammonium molybdate) the magnesium ammonium phosphate is filtered off, washed with ammonia water (1 part of ·880 ammonia to 3 parts of water), dried, and then ignited together with the filter paper. The ignition should be carried out first over the Bunsen flame, and then for five or ten minutes over the blowpipe. The residue is $Mg_2P_2O_7$; 1 part of this equals 0·6379 part of $P_2O_5$, or 1·3907 parts $Ca_3(PO_4)_2$.

[1] *Magnesia Mixture* is prepared as follows:—80 grams of crystallised $MgSO_4$ and 160 grams of ammonium chloride are dissolved in about 600 c.c. of water and then mixed with 120 c.c. of strong ammonia solution (0·880). The liquid is diluted to 1 litre, allowed to stand for some time, and then decanted into a clean bottle.

(*b*) "*Reverted*" *or* "*Citrate Soluble Phosphate.*"—This value is found by difference from the "total phosphates" (see (*d*) below) and the sum of the "water soluble phosphate" (*a*) and "citrate insoluble phosphate" (*c*).

The estimation is rather indefinite, owing to the necessarily arbitrary choice of solvent, different materials being used in different methods. The differentiating solvents employed to distinguish between the various calcium phosphates are: ammonium citrate in neutral, alkaline (ammonia) or acid (citric) solution; 2 per cent. citric acid solution and 1 per cent. citric acid solution. The last named is generally adopted for "available" phosphates in soil (**87,** p 103), the 2 per cent. citric acid for basic slag (**124** (*b*), p. 136), and the neutral ammonium citrate for reverted phosphate in superphosphate.

**117**. *The Ammonium Citrate Solution* is prepared as follows:—370 grams of citric acid are dissolved in 1500 c.c. of water, nearly neutralised with ammonium carbonate, and the carbon dioxide expelled by heat. The solution is then cooled, *exactly* neutralised with ammonia solution, and diluted to 2 litres. The specific gravity of this solution should be 1·09 at 20° C.

(*c*) *Citrate Insoluble Phosphate.*—The residue from (*a*) is washed with water into a flask of 200 c.c. capacity and mixed with 100 c.c. of the ammonium-citrate solution (see above). The flask is corked and heated on the water-bath for thirty minutes at 65° C., the contents being well shaken at frequent intervals. The liquid is quickly filtered, the residue well washed with water, and the filter and contents transferred to a platinum basin, dried, and ignited until all the organic matter is destroyed. The residue is digested with 10 c.c. of concentrated HCl to dissolve the residual phosphate, diluted with water to 200 c.c., filtered, and the phosphate estimated in the filtrate as in (*a*).

(*d*) *Total Phosphate.*—Two grams of the superphosphate are added to 30 c.c. of concentrated hydrochloric acid and heated. About 0·5 gram of potassium chlorate is then added and the mixture boiled gently until all the phosphates are dissolved and the organic matter destroyed. The solution is diluted to 250 c.c. and filtered. Fifty c.c. are abstracted, neutralised with ammonia, and mixed with 15 grams of ammonium nitrate. The phosphate is then precipitated with ammonium molybdate and estimated in the usual way.

**118** *Calcium.*

The whole of the calcium in an aliquot part of the original solution prepared for the water soluble phosphate (**116** (*a*), p. 130) is precipitated as oxalate by ammonium oxalate and weighed as CaO after ignition (see **83**, p. 101).

**119.** *Alkali Salts.*

The residue from estimation of the organic matter (**114**) is extracted thoroughly with water, the solution evaporated, if necessary, to about 150 c.c., and heated on the water-bath to 100° C. Barium-chloride solution is added drop by drop in slight excess, and then, without filtering, baryta water, also in slight excess. The liquid is again heated, filtered, and the precipitate washed. To the filtrate and washings is added 1 c.c. of strong ammonia, then a strong solution of ammonium carbonate, and finally, to the hot liquid, 1 gram of powdered ammonium oxalate. The liquid is allowed to stand for some time, the precipitate filtered and washed, and the filtrate evaporated down to dryness in a platinum basin and ignited carefully until free from ammonium salts. The potassium is then estimated in the residual alkaline chlorides by means of platinic chloride (**79**, p. 99). The sodium, if required, may be obtained by difference from the weight of mixed chlorides.

**120.** *Sulphate.*

The sulphuric acid is estimated in the original solution as $BaSO_4$, by precipitation with barium chloride.

**121.** *Statement of Results.*

Whatever the form of phosphate estimated in agricultural analysis, the result is always stated in terms of the tricalcium phosphate, $Ca_3(PO_4)_2$, or as phosphoric anhydride, $P_2O_5$. One part of the latter equals 2·18 of the former. The results of a complete analysis of superphosphate should be stated as follows:—

| | Per cent. |
|---|---|
| *Moisture* | 20·53 |
| *Combined Water and Organic Matter* containing nitrogen = 0·32 ammonia | 14·76 |
| *Soluble Phosphate* as $Ca_3(PO_4)_2$, containing 11·99 per cent. $P_2O_5$ | 26·14 |
| *Insoluble Phosphate* | 1·81 |
| *Reverted Phosphate* | — |
| *Calcium Sulphate*, $CaSO_4$ | 37·28 |
| *Potash and Soda*, $K_2O + Na_2O$ | 1·48 |
| *Silicious Matter*, etc. | 4·44 |

## Basic Slag.

Basic slag is a by-product in the manufacture of steel from pig-iron which has been prepared from iron ores containing phosphorus. It depends for its activity chiefly on its fine state of subdivision, and 80 to 90 per cent. of a sample should pass the sieve containing 100 meshes to the linear inch. The active phosphate is insoluble in water, but dissolves in a 2 per cent. solution of citric acid. Fully 90 per cent. of the total phosphoric acid should be thus soluble. Basic slag contains a proportion of free lime and also contains sulphides, a certain amount of carbonates, free iron, and magnetic iron oxide. It is sometimes adulterated with ground mineral phosphate.

### 122. *Qualitative Examination of Basic Slag.*

(*a*) *Fineness.*—Ten grams of the sample are sifted through a sieve containing 100 meshes to the linear inch (10,000 to the square inch). When no more comes through, the residue is weighed, the weight subtracted from the total weight, and the percentage fineness calculated. This figure should not be below 80.

(*b*) *Carbonates, Sulphides, Fluorides.*—The slag is tested with diluted hydrochloric acid. If any effervescence takes place the issuing gas is tested for carbon dioxide (lime water) and hydrogen sulphide (lead acetate paper).

Fluorides are tested for in the usual manner after treating the slag with strong sulphuric acid.

(*c*) *Free Iron and Magnetic Iron Oxide.*—The presence of these substances is readily shown by their adhering to a magnet passed through a sample of basic slag.

(*d*) *Lime.*—A few grams of the slag are shaken with 50 c.c. of distilled water and filtered. The filtrate is distinctly alkaline to indicators, and will require an appreciable number of drops of decinormal acid to neutralise.

### 123. *Adulteration of Basic Slag.*

The fertiliser is occasionally sophisticated or imitations made up with ground mineral phosphates. The difference in density may be used for detection. Mineral phosphate is also shown by the presence of fluoride and the absence of free lime. Much carbonate indicates either an old basic slag or the presence of mineral phosphates. If these qualitative tests are equivocal, recourse must be had to analysis.

**124.** *Analysis of Basic Slag.*

(*a*) *Total Phosphate.*—Two grams of the sample are heated in a round-bottomed hard glass flask with 20 c.c. of concentrated sulphuric acid until white fumes appear. The liquid is cooled, diluted with distilled water, and made up to 500 c.c. in a measuring flask. The liquid is well shaken, filtered, and the phosphate determined in 50 c.c. of the filtrate (=0·2 gram of the slag) by the molybdate method.

(*b*) *Citric Soluble Phosphate.*—Five grams of the slag are shaken in a litre bottle for half an hour with 500 c.c. of a solution containing 10 grams of citric acid. The mixture is filtered and the phosphate determined in 50 c.c. of the filtrate.

(*c*) *Fineness.*—The fineness is determined as above (*a*).

# CHAPTER XIII

## POTASH MANURES

WITH the exception of wood ashes and one or two by-products from the manufacture of glue, etc., nearly all the potash fertilisers are soluble salts, derived before the war from the Stassfurt deposits. Since the war, however, considerable supplies of similar potash salts are being obtained from deposits in Alsace, French kainit being an example. During hostilities, recourse was had to the use of flue dust obtained as a by-product in blast furnaces, but owing to the fact that this material is very variable in composition, and slow in action, its employment will largely cease as a result of the more soluble and richer potash compounds again coming into the market.

Large quantities of potash exist in the felspars and other rocks, but in an insoluble condition, and little success has so far attended the use of such materials either in the raw state or after treatment.

The chief potash manures in use are :—

| | | | |
|---|---|---|---|
| *Kainit* . . . | av. 12·8 | per cent. | $K_2O$, chiefly sulphate |
| *Carnallit* . . | „ 9·8 | „ | $K_2O$, „ chloride |
| *Sulphate of Potash* . | „ 51·0 | „ | $K_2O$ |
| *Muriate of Potash* . | „ 52·4 | „ | $K_2O$ |
| *Potash Manure Salts* . | „ 20-30 | „ | $K_2O$, chiefly chloride |
| *Flue dust* . . | „ 3-13 | „ | $K_2O$ |

The low-grade materials like kainit, carnallit, and the potash manure salts contain large proportions of the sulphate and chloride of magnesium and of common salt. Sulphate of lime is also likely to be present in small quantity.

### 125. *Wood Ashes.*

A small quantity of wood chips or sawdust is incinerated at low temperature. The ash is extracted by boiling with water and the extract filtered. It will be found to be alkaline to litmus. The filtrate is neutralised with dilute HCl (note the effervescence), and tested for potassium with platinic chloride or sodium cobaltinitrite solution.

### 126. *Potash Felspar.*

Some orthoclase or other potash-containing felspar is ground to a fine powder and equal quantities are treated as follows :—

(*a*) Extracted by shaking with 2 per cent. citric acid solution.

(*b*) Extracted by boiling with strong HCl.

(*c*) Moistened with water, fused with seven times its weight of ammonium fluoride, and then extracted with dilute HCl.

In each case the filtered extract is tested for potassium, and an attempt made to obtain some indication of how much potash has been removed by each treatment.

## Examination of Potash Salts.

127. *Qualitative Examination.*

Kainit, sulphate of potash, and muriate of potash are tested qualitatively for calcium, magnesium, sodium, sulphate, and chloride. The amount of insoluble material is to be noted, and an attempt made to gauge roughly the relative quantities of magnesium, sulphate, and chloride. In the case of the sulphate and muriate of potash, this may give some indication of purity without having recourse to analysis.

## Analysis of Potash Salts.

Generally speaking, it is sufficient to estimate the potash only, since the manures are bought and sold on this basis. If there is any evidence of sophistication, however, it may be necessary also to estimate the sulphate and chloride present.

128. *Potash.*

(*a*) *In Muriate of Potash free from Sulphates.*—A weighed portion of the sample (about 5 grams in the case of concentrated muriate of potash, or 10 grams in the case of low-grade muriate) is dissolved in water, the solution filtered if necessary, and made up to 500 c.c. in a measuring flask. Fifty c.c. of the solution are pipetted into a porcelain basin, a few drops of hydrochloric acid added, and also 10 c.c. of a 10 per cent. solution of platinic chloride. After evaporation to a syrup on the water-bath, the contents of the basin are allowed to cool, the residue is digested with 80 per cent alcohol (sp. gr. = 0·864) and then transferred to a tared filter paper, or weighed Gooch crucible,

*Note.*—For Perchlorate method of estimating potash, see para. 79.

where washing with diluted alcohol is continued until the washings are colourless. The residue is dried at 100° C. and weighed as $K_2PtCl_6$.

(*b*) *In Salts of Potash containing Sulphates.*—A weighed portion of the sample (5 grams in the case of sulphate of potash, or 10 grams in the case of kainit or other low-grade salts) is boiled with 20 c.c. of HCl and 300 c.c. of water. Barium chloride solution is cautiously added, drop by drop, to the boiling solution until the sulphuric acid is completely precipitated. Any slight excess of barium is removed by the addition of the least possible excess of dilute sulphuric acid. The liquid, without filtration, is cooled and made up to 500 c.c. A portion is then filtered and 25 c.c. of the solution are treated as above in **128** (*a*).

The percentage of potash is returned as $K_2O$.

**129.** *Chloride.*

The chloride present in 10 c.c. of the solution is determined by titration with $N/10$ silver nitrate solution, using potassium chromate as indicator. The chloride is calculated as sodium chloride.

**130.** *Sulphate.*

Twenty c.c. of the solution are precipitated with barium chloride solution in the usual manner, and from the weight of $BaSO_4$ obtained is calculated the percentage of $SO_3$ present.

$$BaSO_4 \times 0{\cdot}343 = SO_3.$$

**131.** *Lime.*

One hundred c.c. of the solution are mixed with 50 c.c. of ammonium chloride solution, and to the liquid is added a small quantity of ammonia and then ammonium oxalate solution in excess. The mixture is boiled for a few minutes and the precipitated calcium oxalate filtered off, washed, dried, and ignited with the filter paper, gently at first and then for fifteen minutes over the blowpipe to CaO. The percentage of lime is calculated as $CaSO_4$.

$$CaO \times 2{\cdot}428 = CaSO_4.$$

# CHAPTER XIV

## Mixed Manures and Calcium Compounds

### Mixed Manures.

MIXED manures, containing two or more of the chief fertilising constituents, may be natural products or artificial mixtures. Of the former class the most important are the bone manures (nitrogen and phosphoric acid), which have already been discussed (**110**, p. 127), fish meals and guanos, true guanos, and the sewage products known as Native Guano, British Guano, etc., which may contain nitrogen, phosphates, and potash.

### Guano.

Three principal types of guano are placed on the market. They include ordinary *Peruvian Guano*, which is largely decayed, and has an average nitrogen content of 5 to 8 per cent., largely derived from uric acid; *Ichaboe Guano*, which is very fresh and contains many undecomposed feathers, etc.; and *Phosphatic Guano*, of the Lobos type, which may contain as much as 60 per cent. of calcium phosphate and only 2 or 3 per cent. of nitrogen.

**132.** *Qualitative Examination of Guanos.*

(*a*) *Peruvian Guano.*

(1) Colour—grey in richer samples; brown in more phosphatic samples.

(2) Smell—ammoniacal.
(3) General Condition—loose, friable powder. May contain fragments of stone and half-decomposed feathers, etc.
(4) Reaction—strongly alkaline.

(*b*) *Ichaboe Guano.*

(1) Colour—grey (see above).
(2) Smell—ammoniacal and excretal.
(3) General Condition—fresh, undecomposed feathers, and sometimes very sandy. High nitrogen content.

(*c*) *Phosphatic Guano.*

(1) Colour—brownish red to dark brown.
(2) Smell—practically none.
(3) General Condition—frequently damp. Low nitrogen content.

(*d*) *Equalised Guano.*—A standard mixture of guano made up by the importers containing about 7 per cent. nitrogen. Characters as in Peruvian guano (*a*).

**133.** *Separation of Uric Acid from Guano.*

Uric acid may be extracted from guano as follows:—A few grams of Peruvian or Ichaboe guano are boiled with dilute caustic soda solution until there is no further smell of ammonia, *i.e.*, all the ammoniacal compounds have been destroyed. The liquid is cooled, filtered, and acidified with dilute nitric acid. The precipitated material is filtered off, washed with alcohol, and a small portion tested for uric acid by the murexide test (see **103**, p. 120).

**134.** *Examination of other Mixed Manures.*

(*a*) *Fish Meal and Fish Guano.*—The following characteristics should be noticed:—

(1) Oiliness—shake a few grams with ether, and note whether much oil is left after filtering and evaporating the solvent.
(2) Condition—the manure should be finely divided, must not contain much sand or shell, and be free from larger stones, etc.
(3) Smell—distinctive fishy odour and well decomposed.

The manure should be tested for ammoniacal and organic nitrogen and nitrates, and for phosphates and potash in the usual manner.

(*b*) *Rape Dust.*

(1) Oil—tested for as above.
(2) Potash and Phosphates—rape dust frequently contains appreciable quantities of these constituents, which may be detected in the usual way after destroying the organic matter.

(*c*) *Sewage Sludges*, etc.—These should be examined as to their condition—fineness, etc., and for organic matter by estimating the loss on ignition, which in some cases is very small.

**135.** *Artificial Mixtures—Incompatible Manures.*

In making up mixtures of artificial manures it must be remembered that certain fertilisers are incompatible owing to the reactions which are set up, possibly causing the loss of some valuable constituent.

(*a*) *Superphosphate and Nitrates.*—A few grams of superphosphate are mixed with a little nitrate of soda or nitrate of lime and the mixture allowed to stand overnight. Nitric acid is liberated in the free condition. The reaction is hastened and made more evident by the action of heat.

(*b*) *Superphosphate and Kainit or Muriate of Potash.*—The experiment (*a*) is repeated with kainit or muriate of potash instead of nitrate. After some time hydrochloric acid is liberated as a gas, and may be detected by holding a piece of blue litmus paper over the mixture.

(*c*) *Lime and Ammoniacal Compounds.*—Sulphate of ammonia, guano, or other fertilisers containing ammonium salts when mixed with lime and slightly moistened quickly give off free ammonia. The reaction is hastened by gentle heating.

(*d*) *Basic Slag and Ammoniacal Compounds.*—Owing to the free lime which it contains, basic slag acts as in (*c*), with the liberation of ammonia from ammonium salts, etc. This can be shown in the same way.

(*e*) *Basic Slag and Nitrates.*—Although without action on each other it is inadvisable to make a mixture of these two manures, owing to the high density of the slag, which prevents proper mixing.

Equal parts of basic slag and nitrate are mixed together and shaken in a closed tube for some time. The slag gradually accumulates at the bottom of the tube.

### 136. *Analysis of Mixed Manures.*

(*a*) *Sampling.*—The sample must be disintegrated as far as possible, pieces of stone, etc., removed and their weight determined, and the residue thoroughly mixed and a sufficient portion withdrawn and ground to a powder.

(*b*) *General Analysis.*—Moisture, nitrogen, nitrates

and water soluble, citric soluble and total phosphates, are determined by the methods previously described. The total phosphates are estimated after destroying organic matter by ignition and rendering the silica insoluble in the usual way.

(*c*) *Potash.*—Ten grams of the sample are gently ignited to destroy the organic matter, heated for ten minutes with 10 c.c. of concentrated hydrochloric acid, and then boiled with 300 c.c. of water. The liquid is filtered, boiled, and a slight excess of baryta water added. The contents of the vessel are cooled and diluted to 500 c.c. in a measuring flask, and then filtered. Two hundred and fifty c.c. of the filtrate are made alkaline with ammonia and treated with excess of ammonium carbonate solution, and then, while boiling, with a little powdered ammonium oxalate. The liquid is again cooled, made up to 500 c.c., and filtered. Of the filtrate, 100 c.c. are evaporated in a platinum dish and the residue heated gently over a low flame until all the ammonium salts are expelled. The residue is then treated with hot water, filtered if necessary, and the potash determined by precipitation with perchloric acid in the usual way.

## Lime and Limestone.

Quicklime, slaked lime, and ground limestone (chalk) are all used as manures, although not in themselves plant foods. They act indirectly, either by preventing acidity, bettering the soil texture, or by liberating the soil reserves of plant nutrients, especially potash.

### 137. *Limestone.*

The calcium carbonate present is determined by the method already described (**36** (*b*), p. 57). Weight of $CO_2 \times 2{\cdot}2727 =$ weight of $CaCO_3$.

A small quantity is dissolved in hydrochloric acid and

the amount of insoluble matter noted; if large, it should be estimated.

Magnesia is tested for qualitatively, and if present in large amount must be determined quantitatively. Good agricultural limestone or lime should contain very little magnesia.

**138.** *Lime.*

(*a*) *Quicklime.*—Examined qualitatively, it should be white, swell considerably on slaking, and fall into a fine powder. Grey limes which do not swell and which slake badly are less valuable for agricultural purposes.

The lime should be examined for insoluble matter and magnesia, as in **137**.

(*b*) *Slaked Lime.*—This material should be in a fine state of subdivision and free from grit, insoluble matter, and much magnesia.

**139.** *Analysis of Agricultural Limes.*

The most satisfactory method of analysing a lime which is comparatively free from magnesia and may contain quicklime, slaked lime, and carbonate, is as follows:—

(*a*) *Insoluble Matter.*—Two or 3 grams are dissolved in HCl, the insoluble material filtered off, and weighed.

(*b*) *Carbonate.*—Estimated by any of the usual methods (**36** (*b*), p. 57, or **75**, p. 94).

(*c*) *Total Calcium.*—Estimated in the usual manner by precipitation as oxalate and weighing as CaO (**83**, p. 101).

(*d*) *Lime.*—The lime present as oxide or hydrate (quick or slaked lime) is obtained by difference from the total calcium, calculated as CaO, and the carbonate also calculated back from $CaCO_3$ to CaO.

(*e*) *Quicklime.* — This may be determined by exposing a weighed sample in an atmosphere of aqueous vapour until all the lime is slaked. After drying at 100° C., the increase in weight is determined and the calcium oxide originally present calculated from the equation $CaO + H_2O = Ca(OH)_2$.

The *slaked lime* in the sample is (*d*)-(*e*).

# SECTION IV.—FEEDING STUFFS

## CHAPTER XV

### COMPOSITION OF FEEDING STUFFS

THE feeding stuffs used for stock vary considerably in composition and are usually very diverse mixtures. They almost invariably contain thirty to forty different substances, many of them, of course, in small amount only. Some of these components are of the highest importance in animal nutrition; many (and, of these, generally the substances occurring in smallest quantity) have no feeding value, as far as our present knowledge goes.

The components of the feeding stuffs, whether of animal or vegetable origin, are consequently classified according to their particular nutritive values, which frequently coincide with their chemical grouping. *Section I.* has dealt fully with the general composition of the plant and its products; it will be sufficient therefore to indicate forthwith the determinations which are made in the chemical examination of a food-stuff. They are:—

(1) Water.
(2) Protein or albuminoids.
(3) Non-protein nitrogenous substances.
(4) Fat or oil. (Ether extract.)
(5) Carbohydrates or nitrogen-free extract.
(6) Crude fibre.
(7) Ash and sand.

According to the particular type of food-stuff being examined, these determinations may be modified or some of them left out. In special cases also, particular estimations are included, as, for example, in the estimation of the gluten of wheat or the sugar in roots.

## Analysis of Feeding Stuffs.

**140.** (*a*) *Moisture.*

Moisture is that portion of a food which disappears on drying at 100° C. In some cases the loss includes, besides water, small quantities of volatile acids and essential oils, but unless for some special reason, no correction is applied for this slight error.

*Preparation of Sample.*—The material is reduced, if possible, to a fine powder and evenly mixed. If the food-stuff is very moist, as in the case of green fodders and roots, the material is chopped finely and the estimation then carried out, the substance being reduced to a powder subsequently.

*Estimation of Moisture.*—Five to 10 grams of the material are heated in the steam-oven to constant weight. The loss is moisture.

In the case of substances such as linseed, which contain large quantities of oil which is found to oxidise readily at the temperature of the oven, the material should be dried *in vacuo*, or in a current of hydrogen or coal-gas, in apparatus of the type shown in Fig. 20.

**141.** (*b*) *Ash and Sand.*

Five grams of the material are ignited to whiteness at a dull-red heat.

*Phosphate.*—Phosphate in the ash, when required, is determined by extracting with HCl to which a little nitric acid has been added, filtering off the insoluble matter, and then estimating the $P_2O_5$ with ammonium molybdate in the usual way.

*Sand.*—The "crude ash" should be examined qualitatively for fine earth, sand, gypsum, chalk, etc., which may have been used as adulterants. The insoluble matter remaining after extracting the ash for "phosphate" is filtered off, ignited, and weighed as "sand or silicious matter."

**142.** (*c*) *Crude Fat or Ether Extract.*

The fat or oil is extracted from 10 grams of the material by the method described in **12**, p. 19, with ether or petroleum ether boiling below 50° C. If the

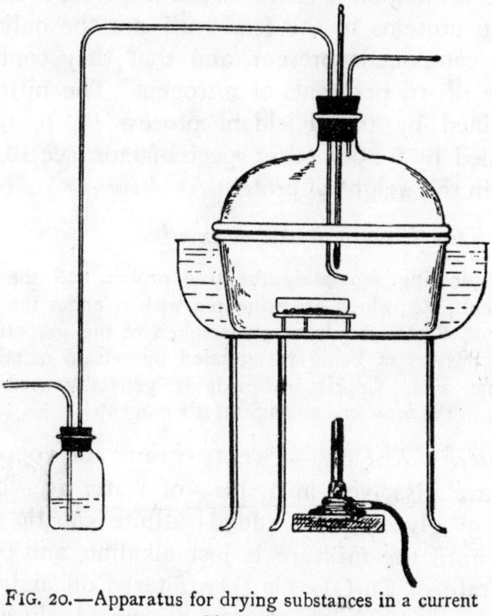

FIG. 20.—Apparatus for drying substances in a current of inert gas.

former is used, it must be dry and free from alcohol, and the sample must be dry also. If petroleum is used the absence of moisture is not so essential.

The dry material from the moisture estimation may be used for the estimation of the oil or fat.

By this method, unfortunately, not only is the fat extracted but also such substances as waxes, colouring matters, organic acids, etc. The separation of these substances from the fat is not easy, and they are therefore included under the heading of crude fat or ether extract.

### Nitrogenous Constituents.

**143.** (*d*) *Crude Protein.*

This estimation is based on the incorrect assumption that the proteins in the food-stuff are the only nitrogenous compounds present, and that they contain an average of 16 per cent. of nitrogen. The nitrogen is determined by the Kjeldahl process (**7**, p. 10) and multiplied by 6·25 or other special factor (see **10**, p. 18), to obtain the weight of protein.

(*e*) *True Protein or Albuminoids.*

In estimating separately the true protein and the amides, amino acids, etc., which are estimated with it under the heading of "Crude Protein," advantage is taken of the property which proteins possess of being precipitated by certain metallic salts or tannic acid. Cupric hydroxide is generally used for the purpose, in the form known as Stutzer's reagent.

*Stutzer's Reagent.*—Twenty grams of copper sulphate are dissolved in 1 litre of water to which 25 grams of glycerol are added; dilute caustic soda is added until the mixture is just alkaline, and the blue precipitate of $Cu(OH)_2$ is then filtered off and washed with water containing glycerol (5 grams glycerol per litre) till the washings cease to be alkaline. The precipitate is rubbed up in a mortar with water containing 10 per cent. of glycerol to form a uniform gelatinous mass, which can be measured off in a pipette. The amount of $Cu(OH)_2$ per cent. of this liquid is determined.

*Estimation of True Protein.*—One gram of the foodstuff sample is added to 100 c.c. of water and the mixture heated to boiling. Substances containing much starch should be heated on the water-bath for ten minutes. A quantity of the cupric hydroxide reagent containing 0·5 to 0·8 gram $Cu(OH)_2$ is added to the mixture, which is stirred well and allowed to cool. The precipitate, which contains copper and the true proteins, is filtered off, washed, dried, and the nitrogen estimated by Kjeldahl's process. $N \times 6{\cdot}25$ (or the same factor used for the crude protein) = Protein.

To substances such as oilcakes, crushed seeds, etc., which contain phosphates, it is advisable to add a few cubic centimetres of a concentrated alum solution before using the cupric hydroxide reagent. This decomposes alkaline phosphates; otherwise alkali may be set free from the phosphates and cupric phosphate formed, some of the albuminoids being dissolved.

(*f*) *Non-Protein Nitrogenous Material.*

These substances, consisting chiefly of amides and amino-acids, etc. (see **11**, p. 18), are present in considerable quantity in some feeding stuffs, *e.g.* mangolds. Their nutritive value is not as great as that of the proteins, and consists largely in conserving the true proteins.

The non-protein nitrogenous matter is obtained by difference from the crude protein and the true protein estimated as above.

### Non-Nitrogenous Constituents.

**144.** (*g*) *Crude Fibre.*

The residue from the ether extract, or another portion of fat-free substance, is boiled for half an hour with 200 c.c. of 1·25 per cent. sulphuric acid, washed free from acid on a calico filter, and then boiled for another half-hour with 1·25 per cent. caustic soda. The digestion should be carried out in a large flask fitted with a reflux tube, as considerable frothing is liable to take place. After the NaOH digestion the residue is filtered through calico again, washed well with hot water, and then transferred, by washing, into a weighed platinum dish. The water is evaporated off on the water-bath, the residue dried in the oven, and weighed. It is then ignited and weighed again. The difference in weight is the "crude fibre."

(*h*) *Carbohydrates — Non-nitrogenous Extractive Matter.*

These substances, consisting chiefly of sugars and starch, are generally estimated by difference from 100 after all the other determinations are made. If it is desired to estimate them separately, the following methods are adopted.

**145.** *Soluble Carbohydrates or Sugars.*

(1) *Reducing Sugars estimated as Dextrose.*—About 5 grams of substance (less if it contains much sugar) are stirred with water and allowed to stand for some time—say overnight—in about 50 c.c. of cold water. The solution is filtered and the residue washed. The filtrate and washings are made up to 250 c.c.

The dextrose is then estimated polarimetrically (see **17** and **18**, pp. 31-34), or by reduction of Fehling's solution as follows:—25 to 30 c.c. of Fehling's solution are put in a 250 c.c. beaker and diluted with 50 c.c. of boiling water. The liquid is kept in a water-bath containing boiling water, and after six or seven minutes (the liquid being quite clear) 50 c.c. of the sugar-containing solution are added and the mixture maintained in the boiling water for twelve to fourteen minutes. If the blue colour of the liquid be destroyed within a few minutes, the assay must be recommenced with a smaller quantity of saccharine liquid. The precipitated cuprous oxide is rapidly filtered through a weighed Gooch crucible, washed with boiling well-boiled water, and dried. The residue may be weighed direct as $Cu_2O$, ignited to CuO, or dissolved off with nitric acid, and estimated volumetrically or electrolytically.

The factors used to determine the weight of dextrose or cane sugar (see below) from the weight of copper are

| | Glucose. | Cane Sugar (After inversion). |
|---|---|---|
| $Cu_2O$ . . . | 0·5042 | 0·4790 |
| CuO . . . | 0·4535 | 0·4308 |
| Cu . . . . | 0·5634 | 0·5395 |

(2) *Cane Sugar* (*Sucrose*).—The aqueous extract obtained above (1) is purified by treating with excess of basic lead acetate[1] solution, any precipitate filtered off, and the excess of lead in solution removed by passing in $H_2S$. The $H_2S$ in solution is removed by bubbling air through the liquid. The cane sugar is now estimated polarimetrically by inversion (see **19**, p. 33), or the reducing sugars formed determined by means of Fehling's solution as above.

**146.** (*i*) *Starch.*

The insoluble residue from (1) above is hydrolysed to dextrose by boiling for two to three hours with dilute hydrochloric acid (20 c.c. HCl, 200 c.c. water) under a reflux. The digest is cooled, neutralised with sodium carbonate, and the dextrose estimated as above.

$$(C_6H_{10}O_5)_n + nH_2O = n(C_6H_{12}O_6)$$

**147.** (*j*) *Pentosans.*

The pentosans are insoluble carbohydrates and the mother substances of the pentose sugars, such as xylose. Their

[1] See footnote, p. 168.

nutritive value is not identical with that of the other digestible carbohydrates, and they are sometimes estimated separately, advantage being taken of the fact that the pentose sugars when distilled with HCl are converted into furfural, $C_4H_3O.CHO$.

$$C_5H_{10}O_5 - 3H_2O = C_5H_4O_2.$$

Five grams of the material are weighed into a distillation flask fitted with a dropping funnel and connected with a condenser. A mixture of 40 c.c. strong HCl and 60 c.c. of water are added, and the liquid heated. As distillation commences, HCl of the same strength is run in drop by drop to replace the liquid distilled off. The process is continued until the distillate no longer reddens filter paper moistened with aniline acetate solution. The distillate is now transferred to a large flask, neutralised with solid sodium carbonate, and then re-acidified with acetic acid. A solution is made up containing 12 grams of phenyl hydrazine and 7·5 grams glacial acetic acid per 100 c.c., and 10 c.c. of this solution are added to the liquid, which is then gently warmed on the water-bath, shaken well, and allowed to stand overnight at the ordinary temperature. The furfural hydrazone formed is filtered through a Gooch, washed well with water, and the residue, together with any material remaining on the sides of the flask, dissolved in alcohol and transferred to a small weighed dish. The liquid is then evaporated to dryness and the dish again weighed. Weight of hydrazone $\times 0{\cdot}516 =$ weight of furfural. An addition of 0·025 gram is made to correct for solubility. Weight of furfural $\times 2{\cdot}0 =$ weight of pentoses.

# CHAPTER XVI

## CONCENTRATED FOOD-STUFFS—OILCAKES, PULSES, CEREALS, ETC.

### Oilcakes.

**148.** *Qualitative Examination.*

Oilcakes should be subjected to the following qualitative tests:—

(*a*) The cake should be broken up in a mortar, its hardness or other characteristics noted, and an examination made with a lens for the presence of strange husks and seeds, fibre, pieces of wood or other extraneous material, mould, insects, etc.

(*b*) The ground cake is mixed with water in a porcelain dish, stirred with the little finger, and presence of any grit or sand noted.

(*c*) The ground cake mixed with water is warmed on the water-bath and the smell observed. It should be perfectly fresh, and free from mouldiness or rancidity.

(*d*) The cake and water mixture is boiled, allowed to cool, and tested with a solution of iodine in potassium iodide for starch. Most of the oil-seeds contain no free starch, and its presence in any quantity, as *e.g.*, in linseed cake, is strong indication of adulteration with flour or milling by-products, or the presence of immature seeds.

The above tests should be applied to all samples of oilcake which are accessible.

*Formation of Prussic Acid from Linseed Cake.*

Linseed and linseed cake contain a cyanogenetic glucoside (see p. 42) named phaseolunatin, and coexistent with it a specific enzyme. When linseed cake is macerated with water and allowed to stand at blood-heat, hydrocyanic acid is very frequently produced, sometimes in considerable quantity, owing to the hydrolysis of the glucoside by the ferment.

**149.** *Qualitative Examination for HCN.*

A few grams of the cake are soaked with water, and kept, in a corked flask, in a warm place for a few hours. The prussic acid may be detected—

1. By the smell.
2. By suspending over the mixture a piece of filter paper soaked in sodium picrate solution. In the presence of HCN the yellow colour of the paper changes to orange red.
3. By steam distilling (see below) and testing as follows:—The distillate is made alkaline with caustic soda, and two drops each of ferrous sulphate and ferric chloride solutions added. The mixture is just warmed, well shaken, and acidified with a few drops of strong HCl. If hydrocyanic acid is present a blue colour of prussian blue is produced, which in time settles down as a precipitate.

*Estimation of Prussic Acid in Linseed Cake.*

This method may be employed for any prussic-acid-producing feeding stuff (Java beans, sorghum, cassava, etc.). It distinguishes between the total prussic acid which can be formed and the "free," or "available," prussic acid which is actually produced when the substance is soaked in water.

**150.** (*a*) *Total Prussic Acid.*

This figure, representing the amount of cyanogenetic glucoside present in the cake or other feeding stuff, is determined as follows:—A known weight (about 25 grams) of the ground material is re-percolated with hot alcohol in a Soxhlet extractor for twenty-four hours. The solvent is then distilled off from the extract, a little water is added, and then sufficient 10 per cent. sulphuric acid to bring the concentration of the acid roughly to 6 per cent., the total volume being about 50 c.c. The mixture is then distilled until no more prussic acid is liberated, the end of the condenser passing under the surface of dilute caustic soda contained in the receiving flask. The distillate is then titrated, according to Liebig's method, with standard silver nitrate solution. (1 c.c. = 0·001 gram HCN.)

$$2KCN + AgNO_3 = AgCN.KCN + KNO_3$$

The water in the distilling flask is replenished from a tap-funnel as required.

(*b*) "*Free*" *Prussic Acid.*

The prussic acid liberated when the finely ground linseed cake or other feeding stuff is soaked with water is estimated in the following manner:—About 250 c.c. of water are placed in a round-bottomed long-neck flask of about 1½ litres capacity, a few drops of toluene added, and the liquid brought to the experimental temperature by immersion in a thermostat. A weighed quantity of the finely ground cake, varying from 25 to 50 grams according to circumstances, is then introduced into the flask, mixed by giving the contents a shake, and the flask then stoppered with a rubber bung and placed in the thermostat. If the water is added to the cake, instead of *vice versa*, the material is apt to clog and not become thoroughly wetted.

When the action has ceased (twelve hours is sufficient at a temperature of 38° C.), the contents of the flask are steam distilled into saturated sodium bicarbonate solution and the prussic acid content of the distillate determined by titration with $\frac{N}{50}$ iodine solution (1 c.c. $= 0{\cdot}00027$ gram HCN).

$$HCN + I_2 = CNI + HI.$$

This method is much more satisfactory for these distillates than the silver nitrate titration method, since reducing substances, which interfere with the end-point, are apt to be present in the distillates.

### 151. *Crude Fibre in Cotton-seed Cake.*

Several kinds of cotton-seed cake are placed on the market, differing according as the seeds have been husked or not. The principal varieties are :—

| | |
|---|---|
| Decorticated Cake (American) . | . Husk removed. |
| Decorticated Cake (Egyptian) . | . Most of the husk removed. |
| Undecorticated Cake (Egyptian) | . Husk present, but no cotton fibre. |
| Undecorticated Cake (Bombay) | . Husk present together with much cotton fibre. |

Samples of different cotton-seed cakes should be examined qualitatively with regard to their content of husk and fibre. The crude fibre is to be estimated (see p. 151) in decorticated, Egyptian and Bombay cakes and the differences noted.

### 152. *Examination of Rape Cake.*

Rape cake frequently contains considerable quantities of wild mustard seed, which renders it undesirable as a feeding stuff ; it is, consequently, frequently extracted with solvents to remove the oil, and the residue used as a fertiliser (see p. 142).

A few grams of the cake are reduced to a powder in the mill, soaked with water, and allowed to stand in a corked flask in a thermostat or other warm place. It should retain its original mild taste. If, at the end of a few hours, it tastes hot or pungent or makes the eyes water, it should be rejected as a feeding stuff, as mustard seed is present.

## Leguminous Seeds.

**153.** *Lupine Seeds.*

Lupine seeds, which form a valuable food, contain several poisonous and bitter principles of an alkaloidal nature. These alkaloids are volatile, and may be removed by soaking and boiling the seeds.

(*a*) Some lupine seeds are extracted with alcohol in a Soxhlet extractor, the alcohol evaporated off, and the residue taken up by warming with 5 per cent. hydrochloric acid. The extract is filtered, and the acid solution tested for alkaloids with Mayer's reagent and iodine solution (see p. 41).

(*b*) The bitter taste of fresh lupine seeds is noted, and 20-30 grams are then soaked in cold or lukewarm water for twenty-four hours, then boiled for an hour, and finally washed well with cold water, the water being changed every two or three hours, or else the washing carried out in a stream of water. The seeds are now dried, weighed, and again tasted, and also tested as in (*a*) for alkaloidal content. During the process ripe lupine seeds lose 12-20 per cent. of dry matter—mostly nitrogen-free extractive material. Unripe seeds lose up to 30 per cent.

**154.** *Java Beans, Mauritius Beans, etc.*

These beans, the seed of ***Phaseolus lunatus***, are a nutritious feeding stuff, but unfortunately produce considerable quantities of prussic acid on soaking with water, due to the interaction of an enzyme with a cyanogenetic glucoside (see p. 42).

(*a*) *Cyanogenetic Glucoside.*—Some Java beans are tested for prussic acid as in **149**, p. 156. The HCN content may be measured by the processes described under linseed cake (**150**, **151**).

(*b*) *Enzyme.*—A few grams of the beans are ground to a meal and then introduced, a little at a time, into a flask containing 200 to 300 c.c. of boiling water. This process kills the enzyme, and prevents the formation of HCN. The mixture is allowed to cool to blood heat, maintained at that temperature for twelve hours, and then tested for HCN as before. None should be present.

### Cereals.

**155.** *Cereal Starches.*

The common cereals—wheat, barley, oats, maize, rice, etc.—are ground finely and small portions of each of the meals examined microscopically under a high power. The nature of the starch grains in each case should be noted and compared, and sketches made (see p. 34).

**156.** *Examination of Wheat.*

(*a*) *Preparation of Sample.*—The wheat is reduced to a fine flour which will pass through the 0·002 mm. sieve.

(*b*) *Moisture and Ash.*—Ten grams of the sample are dried to constant weight in the steam-oven; the residue is then ignited to whiteness at a low red heat.

(*c*) *Wheat Proteins.*

*Total Protein.*—The nitrogen is estimated by Kjeldahl's method in about 1 gram of flour. Factor = 5·69.

*Crude Gluten.*—The crude gluten or water-insoluble protein of wheat is a mixture of at least two distinct substances—the alkali-soluble *gluten* and the alcohol-soluble *gliadin.*

Ten grams of the flour are triturated in a mortar with about 2 c.c. of water. The dough produced should leave the mortar with no trace of adhering. The mixture is allowed to stand for an hour, placed in a fine linen cloth, tied up, and kneaded gently in a stream of water until the washings are clear and consequently free from starch. This may be done by hand or by means of an automatic stamping arrangement. The crude gluten is removed from the cloth, transferred to a weighed dish, dried slowly at 110° C., and weighed.

The elasticity, colour, and other physical characteristics of the crude gluten are noted.

*Gliadin.*—Four grams of the flour are extracted with 100 c.c. of alcohol of sp. gr. 0·8724 (70 per cent. by weight) in a tight-stoppered bottle. The liquid is allowed to stand for three to four hours, with occasional shaking, and is then filtered. Fifty c.c. of the filtrate are evaporated to dryness in a Kjeldahl flask and the nitrogen then estimated in the usual manner. Gliadin $= N \times 5{\cdot}67$.

The milling qualities of wheat and the baking qualities of flour are largely determined by the nature, amount, and composition of the crude gluten. In the average, wheat contains about 11 per cent. crude gluten, but in the hard durum or macaroni wheats it may reach 17 per cent. or more.

From a good wheat, suitable for bread-making, the gluten is elastic and light yellow in colour. Poor glutens from inferior or damaged wheat are sticky and adhere to the cloth, are with difficulty united to a single mass, and are much darker coloured than that from good flour. These variations are partly due to differences in the proportion of gliadin present in the crude gluten; it is stated that the amount of gliadin profoundly affects the baking quality of wheat.

## Barley and Malt.

**157.** *Barley.*

(*a*) Barley is grown chiefly for the preparation of

malt, and the grain should be subjected to the following physical examination :—

(i.) One hundred grains of the barley are counted out and weighed. They should weigh between 4 and 5 grams.

(ii.) A number of the grains are cut across. The surface should be white and starchy. Samples showing a large proportion of "flinty" grains are unsuitable for malting.

(iii.) The colour is noted. It should be uniform, and of a pale yellow or straw colour. The grains should have a thin wrinkled skin.

(iv.) The general condition is noted. The barley should be free from broken grains and have no musty smell.

(*b*) *Chemical Examination.*—The moisture and proteins are determined. The former should not be more than 14 to 15 per cent., and the proteins in a good malting barley not more than 10·5 per cent.

**158.** *Malt.*

(*a*) Good malt should be of a yellow colour, occasionally speckled with brown, and the interior should be pure white. Malt should have a sweet taste; a rapid perception of sweetness in the mouth indicates a sufficient proportion of diastase.

On placing in cold water, malt should float on the surface.

(*b*) *Malt Extract.*—Fifty grams of the finely ground malt are treated in a beaker with 250 c.c. of warm distilled water at 50° to 52° C. The mixture is placed on the water-bath and maintained at this temperature for a quarter of an hour, after which heat is applied until the temperature reaches 60° C. The mash is maintained in this condition until a drop of the liquid gives no blue coloration with iodine solution, *i.e.*, all the starch has been hydrolysed by the diastase of the malt. The temperature is now raised to 70° C. for a short time and the water in the bath then boiled for five minutes. This step should be arrived at about one hundred minutes after the commencement of the operation. The beaker is cooled and the contents filtered. The residue is well washed and the filtrates and washings made up to 500 c.c. An aliquot part (say 50 c.c.) of the solution is evaporated to dryness in a flat-bottomed weighed dish and the residue weighed.

The weight of malt extractive matter should average 70 per cent. of the weight of malt from which it is derived, or about 235 lbs. per quarter.

The solution of the malt infusion or "wort" should be of a light colour, the taste peculiar and sweet and the odour pleasant, somewhat like that of new bread.

**159**. (*c*) *Diastase in Malt Extract.*—About 100 grams of ground malt are soaked in 100 c.c. water at 35° C. for a few hours; the solid matter is filtered off, and the extract mixed with alcohol until no more precipitate is formed. The latter is filtered off, washed with alcohol, and dried in a desiccator. It consists of impure diastase.

A thin mucilage of starch is prepared in the usual way and mixed with a little of the diastase, the mixture then being retained at 60° to 65° C. on the water-bath. From time to time a few drops of the solution are removed and tested with iodide solution. As the experiment proceeds the colour struck with iodine will pass from blue to purple, then to reddish brown, and eventually no colour is formed at all. This implies that all the starch and also the erythrodextrin is destroyed.

Tested with Fehling's solution the liquid will now show the presence of a reducing sugar (maltose).

The above process, suitably standardised, is used for determining the relative diastatic power of malt and malt extract. Under standard conditions and with a mucilage of 0·1 gram starch in 100 c.c., a good extract will cease to give a blue colour with iodine in half an hour at 60° C.

### 160. *"Diastatic Capacity" of Cereals.*

Under the name "diastatic capacity" Wood expresses a fermentation method of determining the cereal sugar, together with that produced by diastatic action.

*Procedure.*—Twenty grams of the flour or finely ground cereal are made into a cream with 20 c.c. of water in a wide-mouthed bottle. Half a gram of yeast is added and well mixed with the liquid. The bottle is then fitted with a stopper through which passes a delivery tube connected with a gas-measuring tube. The bottle is placed in a thermostat at 35° C. and the rate of evolution of carbon dioxide measured.

*Wheat.*—The "diastatic capacity" of a strong and a weak wheat flour should be compared. The strength of a wheat is partially measured by the evolution of carbon dioxide, both as to rate and total amount.

*Barley and Malt.*—The increased rate of evolution and amount of $CO_2$ formed from barley after malting can readily be demonstrated by the above process, thus comparing the diastatic capacities of barley and malt.

### *Rice.*

Rice is composed chiefly of starch, together with about 8 per cent. of protein—rice-gluten. Rice prepared for domestic purposes is highly polished; the outer portion removed during the process (rice polish) is more nutritious than the grain itself, as it contains most of the "oil" and a higher proportion of proteins. Most of the vitamines have also been shown to be removed from the grain by polishing.

**161.** *Rice Starch and Rice Gluten.*

In the case of rice, bruising and washing are insufficient to remove the starch, and it is consequently subjected to treatment with caustic soda.

About 50 grams of rice are ground to a powder and soaked for twenty-four hours in dilute caustic soda, with occasional vigorous shaking. The grain softens and swells, and a proportion of the protein is dissolved out. The supernatant liquid is drawn off and the residual starch brushed through fine sieves to remove husk, extraneous matter, etc., and well washed with water. The grains should be examined under the microscope.

The alkaline liquid is just neutralised with dilute hydrochloric acid, and the precipitated protein filtered off, dried, and ground. It consists of rice gluten, which is sometimes used as a feeding stuff. Its protein nature may be shown by the usual tests (see **9**, p. 15).

**162.** *Maize.*

Maize contains from 5 to 7 per cent. of oil—a much higher proportion than most of the other cereals. One hundred grams of ground maize are extracted with ether in a Soxhlet, the solvent evaporated off, and the amount and nature of the fatty material noted. The characteristic tests for oils may be applied. It should be noted that maize oil is not a drying oil.

**163.** *Cereal By-Products.*

The following cereal by-products should be examined, and their general characteristics noted for identification purposes:—

*Wheat.*—Bran, pollards, middlings, sharps.

*Barley.*—Brewers' grains (wet and dried), malt coombs, barley meal.

*Maize.*—Maize germ meal, maize bran, gluten feed.

*Rice.*—Rice hulls, rice meal or bran, rice polish.

Commercial samples should be examined and compared, if possible, with authenticated genuine samples. Many of the feeding meals and milling by-products are mixed with sweepings, earth, sand, weed seeds, rust spores, chaff, etc., and occasionally are deliberately adulterated with such substances as chalk and gypsum, earth nut, coffee and rice hulls, dried potato pulp, etc. Indeed, these materials are amongst the most sophisticated of the feeding stuffs. A determination and examination of the ash, and observation under the lens and microscope are generally sufficient to detect gross adulteration.

# CHAPTER XVII

## ROOTS, GREEN FODDERS, ETC.

### Roots.

THE root crops are characterised by their high water-content and, in the dry matter, of a large proportion of sugar and non-protein nitrogenous matter. Owing to their physical condition, numerous modifications are necessary in the methods of their chemical examination.

**164.** *General Analysis of Roots—Swedes, Mangolds, Sugar Beet.*

(*a*) *Sampling.*

This should be done from at least a dozen roots, cores being removed from each, or else a vertical section, half an inch wide, passing through the centre.

(*b*) *Water.*

Thin slices from the mixed cores or sections are

dried in the air-oven at 50° C. for about forty-eight hours. The loss in weight is water.

(*c*) *Nitrogenous Substances.*

*Crude Protein.*—Estimated in a portion of the dry matter by Kjeldahl's method, using salicylic acid (see p. 17).

*Non-Protein.*—Consisting of nitrates, amides, amino acids, etc., is estimated by difference from true protein, which is determined as in **143** (*e*), p. 150.

(*d*) *Sugar.*

Sugar is determined in the pulped sample by extraction of 10 grams, as in **145**, p. 152, or else by the method described below for sugar beet (**168**).

### *Determination of Sugar in Sugar Beet.*

Cones or sections are removed from the beets as in **164** (*a*), after removing the leaves. The portions removed are rasped with a hand-grater, or in a machine with rotating knives.

The determinations made on the juice are :—

(1) Density, by Brix hydrometer.
(2) Sugar percentage by polarimeter.

The data thus obtained give, both directly and by calculation, figures for (*a*) Total Solids in Juice; (*b*) Sugar in Juice; (*c*) Non-sugar in Juice; (*d*) Coefficient of Purity; (*e*) Sugar in Beet.

### **165.** *Expression of the Juice.*

The pulp is enclosed in calico and expressed in a strong hand-press.

*Total Solids in Juice.*—The juice obtained by expression as above is allowed to stand for one hour in order to remove all air-bubbles, and is then transferred to a tall cylinder, and its density measured by a Brix hydrometer. It is usually found necessary to destroy the froth on the surface with a few drops of ether.

The Brix instrument is graduated to give, in direct percentages, the solids in solution (the readings being corrected for temperature according to a scale on the bulb).

**166.** *Sugar in Juice.*

This is estimated in a polarimeter, which should be graduated in sugar degrees, *i.e.*, in such a manner that the percentage of cane sugar can be read off without calculation. According to the type of instrument, there is a normal weight of sugar which should be used in a 200 mm. tube for the graduation.

The volume containing the treble normal weight (16·19 for a Laurent instrument) of the juice is calculated or ascertained from tables. This volume, *e.g.* (containing 48·57 grams), is measured from a burette into a 100 c.c. measuring flask. An Erdmann float should be used in the burette. To the flask is added 5 c.c. of basic lead acetate solution,[1] the mixture well shaken, and a little ether added to dispel froth. The flask is then filled up to the mark with distilled water, the solution filtered, and the filtrate polarised in a 200 mm. tube.

The "sugar degrees" reading divided by 3 gives the percentage of sugar in the juice.

If desired, the dextrose present may be determined indirectly by estimating the true cane sugar after inversion (see **19**, p. 33), or by reduction of Fehling's

[1] Basic lead acetate solution is the most satisfactory clarifying agent known, and is prepared by boiling 264 grams of yellow litharge with 464 grams of neutral lead acetate in 1000 c.c. of water for half an hour. The liquid is then allowed to cool, diluted to 2 litres, left to settle, and decanted.

solution, but it is generally sufficiently accurate to regard all optically active substances, after clarification with basic lead acetate, as cane sugar.

**167.** *Purity of Juice.*

The purity of the juice, or the "coefficient" or "quotient" of purity, is the percentage of sugar in the solids of the juice, and is obtained by multiplying the percentage of sugar in the juice by 100 and dividing by the percentage of total solids in the same juice.

By many investigators this factor of the "purity" of the juice is regarded as one of the most important figures determinable in beet analysis.

**168.** *Sugar in Beet.*

From the percentage of sugar in the juice the amount of sugar in the beet is determined by means of a factor. This factor is a correction for the amount of "marc" or cellular matter existing in the beet, and is liable to a certain amount of variation. The beet may be taken as 95 per cent. juice and 5 per cent. "marc," and the factor should therefore be 0·95.

## Silage.

**169.** *Examination of Silage—Detection of Organic Acids, etc.*

Silage is sour fodder prepared by preserving such substances as the grasses, maize, root-tops, etc., in a closely compacted condition out of access of air. The principal fermentations occurring are the acetic, lactic, and butyric, the particular acids named being produced. A small amount of alcoholic fermentation also goes on.

Several hundred grams of silage are shredded as finely as possible, digested with an equal weight of cold water, and the mixture expressed. The filtrate reacts acid to the usual indicators. The liquid extract is now distilled until the distillate is no longer acid.

(*a*) *Residue.*—The residual material is filtered and shown to be acid to litmus paper. Excess of zinc carbonate is then added, the liquid filtered, and the solution allowed to evaporate. Crusts of zinc lactate separate out. These are recrystallised from water and identified, if desired, by estimating the zinc present. Crystallised zinc lactate contains 18·2 per cent. water of crystallisation and 27·3 per cent. zinc as ZnO.

(*b*) *Distillate.*—The distillate is made alkaline with caustic soda and again distilled. The new distillate now contains any alcohol originally present, together with volatile aldehydes and other products formed in the silo. The first 5 c.c. of distillate are made alkaline with NaOH. and iodine solution added until the reddish tint just remains. The solution is then warmed gently. The presence of alcohol is detected by the formation of iodoform, which may, however, be insufficient to separate out and only be detected by its characteristic smell.

The alkaline liquid in the distilling flask is now acidified with dilute $H_2SO_4$ and again distilled. The distillate will contain any acetic and butyric acids. The latter may be detected by its smell, the acetic acid by the following tests :—

(i.) A few drops of alcohol are added to 2 or 3 c.c. of distillate, and then an equal volume of strong sulphuric acid. On warming gently the characteristic smell of ethyl acetate may be noticed.

(ii.) A small quantity of the distillate is neutralised by adding excess of ammonia and boiling till neutral. The solution is cooled, and mixed with a few drops of ferric chloride. A red colour of ferric acetate is formed, changing to a reddish brown precipitate of the basic acetate on boiling.

During the preparation of silage the green fodder loses a large proportion of its weight. The sugars disappear almost completely and the proteins are lessened considerably, largely by change into amides and amino acids. The changes occurring during the preparation of silage may be followed by analysis conducted in the ordinary way, if representative samples of the green material and the silage are obtained.

# CHAPTER XVIII

## SECONDARY FEEDING STUFFS. DIGESTIBILITY DETERMINATIONS

### Molasses and Molasses Feeds.

MOLASSES is the residual product from the manufacture of sugar from the cane or from the beet. The chief constituent is cane sugar (about 50 per cent.), but dextrose and raffinose are also present. Beet molasses contains a higher proportion of nitrogenous matter than cane molasses ; this consists almost wholly of amides, etc. A large proportion of dissolved salts brings the ash up to 7 or 8 per cent.

Owing to the difficulty of handling, molasses is very frequently absorbed into dry porous material like feeding meals, bran, beet slices, etc., and also into more worthless and indigestible substances like peat moss, spent hops, sawdust, etc.

**170.** *Examination of Molasses.*

(*a*) *Moisture.*—This is an estimation of importance, and requiring a considerable amount of care. About 10 grams of molasses are weighed out accurately, dissolved in water, and the solution made up to 100 c.c.; of this liquid, quantities of 20 c.c. are pipetted into weighed flat-bottomed shallow dishes containing dried fragments of pumice. Dishes and contents are then dried in vacuum at a temperature not exceeding 75° to 80° C.

(*b*) *Insoluble Matter.*—A weighed quantity of molasses is dissolved in water and filtered. The residue, if any, should be examined qualitatively, and if large in amount must be weighed after drying. Coal dust and similar substances are sometimes added to feeding molasses in order to prevent depredation at the farm.

The ash content of beet molasses is much higher than that from canes, the salts sometimes giving the product a peculiar saline taste, and causing scouring in animals which are fed with it.

(*c*) *Ash.*—About 5 grams of the molasses are mixed with a few drops of sulphuric acid and the mixture gently and carefully incinerated. The presence of the acid is of some advantage at the commencement of carbonisation, and renders the process easier of accomplishment. The weight of the ash obtained is diminished by one-tenth to allow for the conversion of the carbonates, etc., into sulphates.

(*d*) *Reducing Sugars.*—The quantity of dextrose in molasses is generally not very large. Reducing sugars may be detected and estimated by the usual methods.

### 171. *Molasses Feeds.*

(*a*) *Absorbent Base.*—About 50 grams of the feed are extracted with warm water and the residue washed until no further soluble material is removed. The residual material is examined under the lens or microscope and identified if possible. Special search should be made for worthless hulls like those of coffee, earthnut or rice, or for sawdust or other woody fibrous material.

The crude fibre in the absorbent base is estimated by the ordinary method.

(*b*) *Molasses.*—10 grams of the material are extracted with warm water as in (*a*), and the residue collected in a Gooch crucible, dried, and weighed. The solution is diluted to 100 c.c., and of this 10 c.c. (equal to 1 gram of the feed) is evaporated for moisture estimation, as in **170** (*a*). The molasses in the remainder of the solution may be examined, if desired, by the processes described above (**170**).

### Condiments, etc.

**172.** *Precipitated Phosphate of Lime.*

This material is sometimes used, under various names, as an adjunct to feeding stuffs.

The material should be examined qualitatively for arsenic and soluble calcium salts and the citrate soluble phosphate estimated as in **124** (*b*), p. 136. Ninety per cent. of the phosphate present should be soluble in citric acid or ammonium citrate, as otherwise there has probably been adulteration with bone meal or bone ash, which can only be assimilated to the extent of 13-14 per cent. by the animal, as against 50-60 per cent. of the true precipitated phosphate of lime.

**173.** *Spices, Cattle Powders, etc.*

Appetisers, conditioning powders, etc., are largely placed on the market. For an examination of the condiments one must rely largely on the sense of smell and the detection of such substances as fennel, aniseed, locust bean, fenugreek, gentian root, etc., of which they are very frequently composed.

Cattle powders sometimes consist merely of common salt, bicarbonate of soda, Glauber's salts, powdered charcoal or sulphur, and such like substances, mixed with a little spice. The detection and estimation of such substances is attended with no difficulty.

## Digestibility of Feeding Stuffs.

The measurement of the digestibility of the various constituents of a feeding stuff is of prime importance, since the animal is only able to utilise that portion of the food which is broken down in the alimentary canal. Ordinary chemical analysis is unable to distinguish between the total food nutrients and those which are digestible, and recourse is had (1) to digestibility trials with animals, or (2) to methods of artificial digestion. In the former method, the fæces of the animal are collected and weighed and the constituents determined as in the food. The second method has only been elaborated with any success in the case of the proteins, which may be digested artificially by means of pepsin.

### 174. *Digestibility of Protein.*

A pepsin solution is prepared by dissolving 1 gram of commercial pepsin, *e.g.* Witte's pepsin, in 1 litre of 0·33 per cent. hydrochloric acid.

The food-stuff (wheat or oilcake will do well for practice) is ground to a fine meal and 2 grams are weighed out and suspended in a flask in 100 c.c. of the pepsin solution. A drop or two of toluene is added to keep the liquid aseptic, the flask is plugged with cotton wool and maintained in a thermostat, with occasional shaking, for twelve hours at 37·5° C. At the end of that time the contents of the flask are poured on a wet filter, the residue washed well with lukewarm water, and then with alcohol and ether. Filter paper and contents are then transferred to a Kjeldahl flask and the protein estimated in the usual way. The total protein is estimated in the food-stuff, and the amount which has been digested is calculated per cent.

# SECTION V.—DAIRY PRODUCTS

## CHAPTER XIX

### MILK

MILK is the natural secretion of special glands of female mammals, and being intended for the nourishment of newly born animals, naturally contains all the nutrients required for their maintenance and growth, namely, protein, fat, carbohydrates, and mineral salts. The proportions of these constituents vary largely in the case of different animals, but in the case of any one species of animal, such as the cow, the percentages do not usually vary between wide limits.

From the point of view of milk-production for domestic purposes, the cow is by far the most important, and in this volume the word milk applies to cows' milk, unless stated definitely to the contrary.

*Average Composition of Cows' Milk.*

| | |
|---|---|
| Water . . . . | 87·75 per cent. |
| Fat . . . . | 3·40 ,, |
| Protein . . . . | 3·50 ,, |
| Milk sugar . . . | 4·60 ,, |
| Mineral substances . . | 0·75 ,, |
| | 100·00 ,, |

The protein in milk consists of casein (85 per cent.), lacto-albumin (15 per cent.), and lacto-globulin (traces), and of these the lacto-albumin is coagulated by heat (70° C.), the casein is precipitated by rennet or dilute acids, while the lacto-globulin is thrown down by the action of mineral salts (*e.g.* $MgSO_4$).

*Colostrum.*—The udder secretion immediately before and after parturition differs markedly from ordinary milk. It is a thick, slimy, yellow liquid, containing about 4.5 per cent. of casein and 15·5 per cent. of albumin, and owing to the large proportion of the latter protein, colostrum coagulates very readily on heating. The fat of colostrum (3·4 per cent.) is poor in volatile acids, and hence gives a low Reichert-Meissl number (p. 200).

## Sampling.

Owing to the fact that the fat globules present in milk are considerably lighter than the milk serum in which they are suspended, it follows that milk which has been standing for some time will be no longer homogeneous, owing to a rising of the fat (cream) to the surface.

Before carrying out any tests or quantitative operations with milk, even if the milk has been standing for only a short time, it is, therefore, necessary that it be thoroughly mixed, in order that the sample obtained may be a representative one. The mixing is conveniently effected by pouring the milk three or four times from one vessel to another, and then immediately withdrawing the sample before the lighter fat globules have had time to rise to the surface.

In the case of milk that has been standing for some time, it is often found that the cream layer does not readily mix with the rest of the liquid. By warming to a temperature of about 40° C., however, it will be found that the process of mixing can be readily performed, and a representative sample thus obtained.

### 175. *Qualitative Examination of Milk.*

(i.) *Albumin.*—About 20 c.c. of milk are heated in a porcelain basin on the water-bath for a short time, and the skin of albumin formed on the surface is removed, washed with water, and examined by the xanthoproteic and biuret reactions (p. 15).

(ii.) *Mineral Ash.*—The liquid remaining in the basin after the removal of the albumin is evaporated to dryness, any skin on the surface being kept broken by stirring with a glass rod. The contents of the dish are then ignited over a Bunsen burner or in the muffle furnace, the carbonaceous residue containing the mineral salts extracted with water, and the aqueous solution filtered and evaporated down to a volume of about 5 c.c.

The solution is then examined qualitatively, according to the method given on p. 9 for plant ash.

(iii.) *Casein.*—About 75 c.c. milk are warmed to a temperature of 50° to 55° C., and allowed to run slowly into 200-300 c.c. dilute $H_2SO_4$ with constant stirring. The casein separates out in the form of a curd, which holds mechanically a proportion of the fat of the milk. To separate this, the liquid is filtered and the filtrate retained for examination for milk sugar (below). The curd is stirred thoroughly with absolute alcohol to remove water, and is then twice shaken up with small quantities of ether. The ether is then filtered off and examined for fat as below. The casein is then examined by the usual tests for proteins (p. 16 *et seq.*).

(iv.) *Fat.* — The ethereal solution of butter-fat obtained by treatment of the casein with ether is evaporated on the water-bath. An oily residue of fat is obtained which will leave a permanent greasy stain on paper.

(v.) *Milk Sugar.*—The acid filtrate from the casein precipitation above is boiled and filtered. A small proportion is then concentrated by boiling, and Fehling's test applied to the liquid. The production of the red precipitate of $Cu_2O$ shows the presence of reducing sugars, glucose, and galactose which have been formed by hydrolysis of lactose or milk sugar. Lactose itself will reduce Fehling's solution.

### The Specific Gravity of Milk.

The specific gravity of pure fresh milk is about 1·032, but undergoes a slight increase of about ·001 during the five hours following milking, this phenomenon being possibly due to some change in the state or condition of the casein.

The determination of the specific gravity of milk is of great importance in milk analysis, and is generally carried out by means of the lactometer or Westphal's balance, although the most scientifically exact method, and that by which the lactometer and Westphal's balance are graduated, is the employment of the density bottle or pyknometer (Figs. 21 and 22).

The specific gravity or density of a sample of fresh milk should be determined by each of the three following methods and the results compared.

### 176. *Determination of Specific Gravity by Density Bottle.*

A density bottle similar to that shown in Fig. 21 which has a capacity of 100 c.c., is washed successively with water, alcohol, and ether, and is then dried internally by *drawing* a current of air through it. The bottle and stopper is then weighed (wt. = $a$ grams).

FIG. 21.—Specific gravity bottle.

It is then filled with cold distilled water, the stopper replaced, and the bottle immersed up to the neck for some minutes in a beaker containing water at 15° C. Any liquid passing through the hole in the stopper is carefully removed, and the bottle is kept in the bath until no further expansion of the water within takes place.

It is then carefully dried externally, care being taken not to warm it with the hand while so doing, and is again weighed (wt. = $b$ grams).

The water is poured out, the bottle thoroughly dried as before, and then filled with milk, immersed in a bath at 15° C., and treated exactly as described above. It is then weighed (wt. = $c$ grams).

Now, weight of water in bottle at 15° C. $=(b-a)$ grams.
" milk " " $=(c-a)$ grams.

Hence specific gravity of milk $= \dfrac{c-a}{b-a}$

If the pyknometer (Fig. 22) be employed, the method adopted is identical in every way, excepting that the liquid is sucked up into the vessel, and the level in the right-hand limb adjusted to the mark *b* by carefully applying a piece of filter paper to the point *a*.

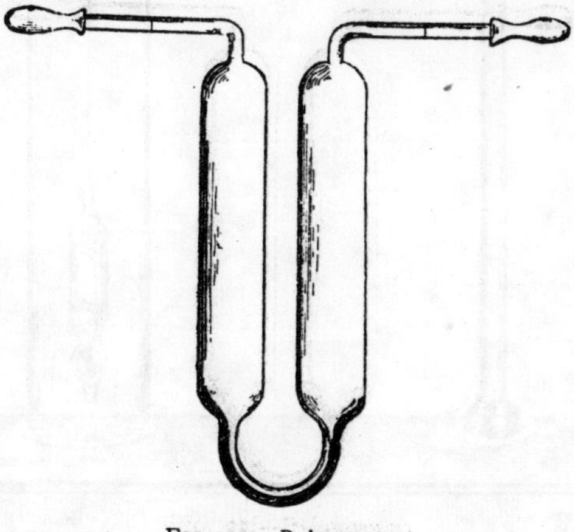

FIG. 22.—Pyknometer.

### 177. *Determination of Specific Gravity by Westphal's Balance.*

Westphal's balance (Fig. 23) is an apparatus for obtaining the densities of liquids by determining the weight required to sink a standard float below the surfaces of the liquids. The weights employed are in the form of wire riders, which can be placed at different positions on the same beam as that to which the float is attached. The density is then read off directly from the graduations on the beam, and in most instruments can be obtained to four places of decimals.

A sample of milk is thoroughly mixed (see p. 176) and is poured into a gas jar or other vessel of sufficient diameter to allow at least a $\frac{1}{4}$ inch clearance between

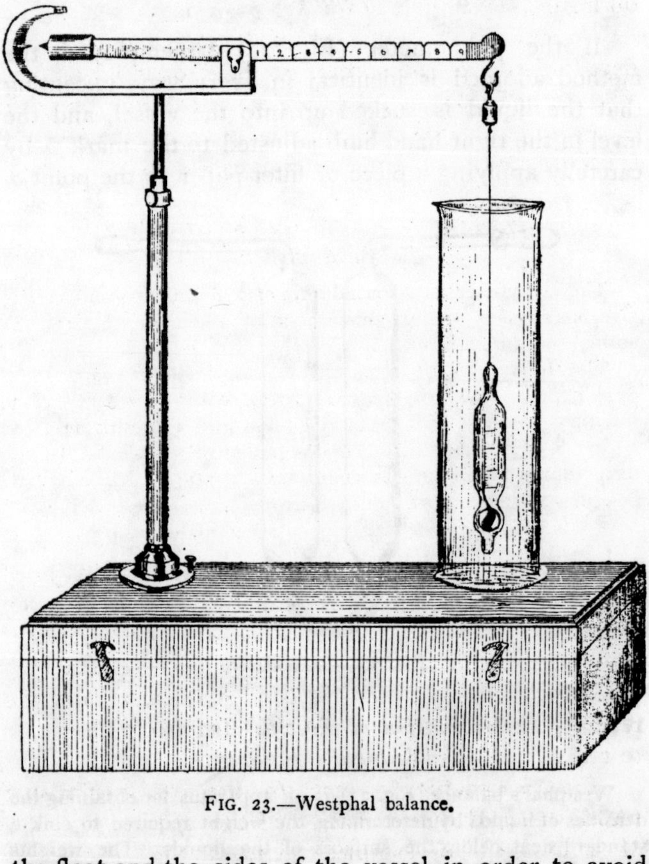


FIG. 23.—Westphal balance.

the float and the sides of the vessel, in order to avoid the effects of surface tension. The level of the surface of the milk must be sufficiently high to allow of a complete immersion of the float. The riders are adjusted, commencing with the largest rider, until the beam is horizontal, as shown by the pointer attached to the short arm, and the density of the milk is then read off from the positions of the various riders.

Westphal's balance allows of the rapid and accurate determination of the specific gravities of milk samples. The method, however, is purely empirical, and the beam is always graduated in the first instance by means of the specific-gravity bottle method (above).

### 178. *Determination of Specific Gravity by Lactometer.*

The lactometer is a special form of hydrometer graduated to give readings in the neighbourhood of a density of 1·032, the normal specific gravity of milk.

The numbers on the lactometer scale represent the second and third numbers after the decimal place, so that a reading of 30 corresponds to a specific gravity of 1·030, and a reading of 42 to 1·042. The lactometer is generally graduated for use at a definite temperature, 15° C., but it is sufficiently accurate to allow a latitude of 5° on each side of the standard temperature, the density at 15° being then determined by means of conversion tables.

A well-mixed sample of milk is poured into a jar of suitable diameter, and the lactometer, which must be clean and dry, is held by the stem, and gently lowered into the milk until the level is somewhere near the reading to be expected. It is then released, and the reading carefully taken when the instrument has come to rest.

The lactometer is easy to use, and affords a means of rapid determination of the specific gravities of a number of samples. Unless the instrument is manufactured by a good maker, however, the results are often not reliable, and should be compared with those obtained by the specific-gravity bottle or pyknometer.

## The Acidity of Milk.

Fresh milk always shows a distinct acidity with regard to phenolphthalein, owing to the presence in it of calcium hydrogen phosphate and dissolved carbon dioxide.

On keeping milk, the acidity is found to increase owing to the production of lactic acid by fermentation of the milk sugar.

The quantitative determination of the acidity of milk is of importance as it affords valuable indication as to how far the lactic acid fermentation has proceeded, and in the case of cream for butter-making, the process of "ripening" can be followed.

The acidity of milk is best determined by direct titration with $\frac{N}{10}$ caustic soda, using phenolphthalein as indicator.

Although it is somewhat difficult to detect the first appearance of the pink coloration owing to the opaqueness of the milk, the latter should not be diluted to remedy this, as the addition of water brings into solution certain alkaline phosphates which have a marked effect in decreasing the acidity.

**179.** *Determination of Acidity.*

Fifty c.c. of fresh milk are placed in a conical flask, a few drops of phenolphthalein added, and directly titrated with $\frac{N}{10}$ caustic-soda solution. In order to facilitate the detection of the pink colour, the titration should be carried out while holding the flask over a piece of white (or cream) paper. As the end-point of the titration is approached, the flask should be inclined after each addition of alkali, as it is easier to detect the coloration in a thick layer of liquid.

The number of c.c. of $\frac{N}{10}$ alkali required multiplied by 2 will give the amount required for 100 c.c. of milk. This number is called the *degree of acidity*. Fresh milk generally has about 18° to 19° of acidity, and this value increases to a marked extent on keeping. To show this, a sample of the same milk should be kept in a loosely stoppered flask and portions of 50 c.c. withdrawn and titrated from time to time, the degrees of acidity being plotted against time since milking.[1]

**180**. *Determination of the Dry Matter or Total Solids.*

The water present in milk is driven off by heating for some time at 100° C., and the substances remaining, consisting of fat, proteins, milk, sugar, and salts, are termed the dry matter or total solids.

Direct evaporation of the milk at 100° C. is not generally employed, as the skin of coagulated albumen formed on the surface interferes with the evaporation. If the skin be repeatedly broken, however, accurate results are obtained.

The most usual method is to employ some absorbent material such as sand or powdered pumice, to which the milk is added before the heating is commenced.

About 12 grams of sand or powdered pumice are ignited, cooled, and placed, together with a glass rod to act as stirrer, in a platinum or porcelain basin, which is then heated to 100° C. in the steam oven for an hour, and weighed.

About 10 c.c. of the milk is placed in a small beaker, which is weighed with its contents; the milk is then poured on to the sand or pumice in the basin, the beaker again weighed, and the weight of milk taken obtained by difference.

The dish is then placed in the steam oven and heated for two or three hours, the mixture being stirred repeatedly. After cooling in the desiccator and weighing, the heating and stirring are again continued for some time, and the basin and contents cooled and weighed as before. The drying is not complete until two successive weighings give the same result. When this is attained the percentage of dry matter left is calculated.

[1] If the degree of acidity be multiplied by ·009, the actual percentage of lactic acid in the milk is obtained.

The amount of total solids in milk may vary from 11 to 14 per cent., the average value being about 12·3.

If the temperature during the heating be much above 100° C., there is a danger of a slight decomposition of the milk sugar occurring.

### Fat.

The fat (butter-fat) existing in milk is a mixture of various glycerides of ten or more different fatty acids, among which are butyric, caproic, caprylic, capric, lauric, stearic, oleic, etc. It is probable that the glycerides are largely mixed tri-glycerides, that is to say, consist of two or three different acid radicles combined with one glyceryl group thus—

$$\begin{array}{ll} CH_2.OOC.C_3H_7 & \text{(radicle of butyric acid).} \\ | & \\ CH.OOC.C_5H_{11} & \text{( " caproic " ).} \\ | & \\ CH_2.OOC.C_9H_{19} & \text{( " capric " ).} \end{array}$$

The fat is present in milk in the form of an emulsion of microscopic globules, and the globules are separate and do not run together except as the result of shaking or other violent mechanical treatment (churning) (p. 196).

The quantity of fat in milk is liable to considerable fluctuations, depending on the breed of cow (Jersey cows give the richest milk), individuality, period of lactation, frequency of milkings, and in separate milkings the values may vary between 1 and 8 per cent. With the mixed milk of a number of different cows, however, the percentage is nearly always between 2·5 and 4·5. Owing to the fact that the value of milk largely depends upon the percentage of fat it contains, and owing to the easy removal of part of the fat or dilution of the milk with water, it is usual for most countries to fix a legal minimum (in England, 3·0 per cent.) for the percentage of fat. Milk falling below this minimum is regarded as having been skimmed or diluted, unless it can be proved to be the untreated and unadulterated product of cows, in which case the prosecution is not proceeded with. The actual determination of the percentage of fat in milk is naturally, therefore, an operation of the greatest importance.

**181.** *Estimation by Adam's Paper-coil Method.*

In this method, which is the standard extraction process, and capable of great accuracy, a certain quantity of milk is soaked up by absorbent paper, which is then dried, and extracted with ether. Special strips of absorbent paper are manufactured for this process, and are the most convenient to use, although blotting or filter paper that has been previously extracted with ether may, however, be employed, after it has been cut into strips about 20 inches long and 2 inches in width.

Ten c.c. of well-mixed milk are run from a pipette upon a strip of ether-extracted paper of the size mentioned above. The paper is then dried by exposing it to the air or holding it at some distance from a fire, the last traces of moisture being driven off by suspending it by means of a piece of platinum wire in a steam-bath at 100° C., for about an hour. It is then rolled round a sinker, consisting of a hermetically sealed and weighted glass tube, which is then placed in a Soxhlet extractor, fitted with a dry flask of known weight, and extracted with dry ether for three or four hours on the water-bath. The sink and paper are then removed, and as much of the ether as possible distilled from the flask into the extractor. The flask containing the fat is dried for two hours in the steam oven at 100° C., cooled in a desiccator, and weighed. On deducting the weight of the empty flask, the weight of the fat is obtained, and from a knowledge of the specific gravity of the milk taken, the percentage of fat in the sample can be calculated.

**182.** *Estimation of Fat by Gerber's Modification of Babcock's Process.*

Of the different methods for fat estimation depending on the mechanical separation of the fat by centrifugal force, the one most generally employed is a modification of the method originally proposed by Gerber in 1892. In order to allow of the complete separation of the fat, it is essential to dissolve the casein present in the milk, and Gerber employed concentrated sulphuric acid for this purpose. At a later date Leffmann and Beam showed that the addition of a small quantity of amyl alcohol facilitates the separation of the fat in a pure clean condition, and this method is the one which is at present most largely employed in dairies, etc., where a rapid method of fat estimation is required.

A Gerber's test-bottle is shown in Fig. 24, and the centrifuge in Figs. 25 and 26. Some of the latter have an arrangement for heating whilst in motion, and may be driven by a strap, bevel-gearing, water-motor, or other means.

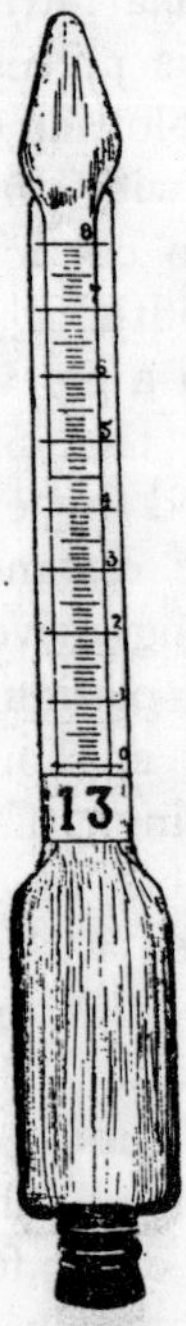
FIG. 24.—Gerber's test-bottle.

The method of carrying out the test is as follows:—

A test-bottle is placed with the open end upwards, and 10 c.c. of pure sulphuric acid of a density of 1·825 at 15° C. are delivered into the bottle, either from a pipette or from a special measurer (Fig. 26). The sulphuric acid should not be allowed to come into contact with the neck of the bottle.

Eleven c.c. of the milk are then run carefully on to the surface of the acid from a special pipette, graduated to deliver this volume, and 1 c.c. of chemically pure amyl alcohol (boiling-point 128° to 130° C.) is added from a 1 c.c. pipette. The reagents should be used in the order mentioned above.

The rubber stopper is fitted into the top of the test-bottle, which is then wrapped in a cloth and thoroughly shaken. A considerable rise in temperature results during the mixing, and the test-bottle may be either immediately placed in the centrifuge, or, in

FIG. 25.—Gerber's centrifuge.

[Face p. 186.

FIG. 26.—Apparatus used in the estimation of milk-fat by Gerber's process.

[Face p. 186.

the case of performing the test with a number of samples, be warmed to 60° to 65° C. for a few minutes in the water-bath (Fig. 27) prior to being centrifuged. When the centrifuge is fitted with a heating arrangement, this preliminary warming is not necessary.

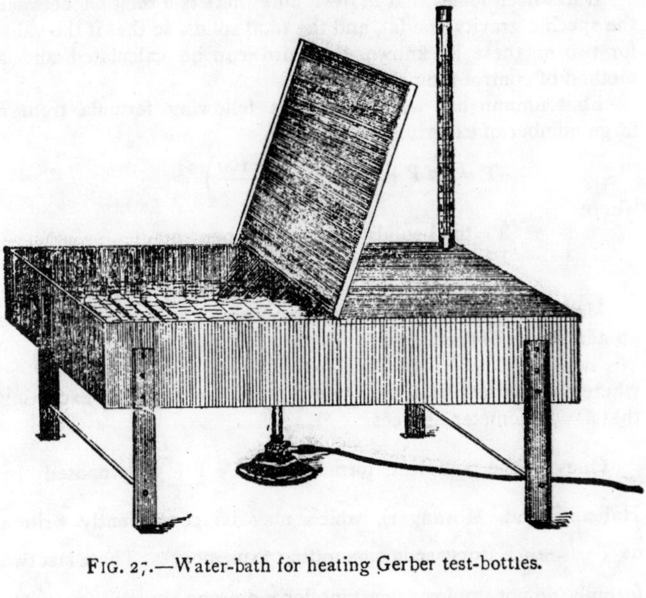

FIG. 27.—Water-bath for heating Gerber test-bottles.

The centrifuge containing one or more test-bottles is then revolved at a speed of about 1000 revolutions per minute for four minutes, the test-bottles removed and placed in the water-bath at 60° to 65° C. for several minutes. By holding the test-bottle in an inverted position and carefully screwing in the rubber stopper, the lower level of the fat column, which is usually of a light yellow colour, is adjusted to the zero mark, and the percentage of fat read off directly on the graduated scale.

Gerber's method, if carefully carried out, gives excellent results comparable with those obtained by the most accurate of the direct extraction methods, such as Adam's method (p. 185).

## Relation between Specific Gravity, the Fat, and Total Solids.

It has been found that in fresh milk there is a relation between the specific gravity, the fat, and the total solids, so that if the value for two of these be known, the third can be calculated and a method of control thus obtained.

Fleischmann has worked out the following formula from a large number of experimental results :—

$$T = 1{\cdot}2\,F + 2{\cdot}665\left(\frac{100\,S - 100}{S}\right)$$

where

T = total solids expressed in percentage,
F = fat expressed in percentage,
S = specific gravity of the milk at 15° C.

Many other formulæ have been proposed, Richmond suggesting an adaptation of the above :

$$T = 1{\cdot}2\,F + {\cdot}25\,D + {\cdot}14,$$

wherein the letters have the same significance as above, excepting that D = lactometer degrees.

One of the simplest formulæ is $T = \frac{F + \frac{D}{5}}{\cdot 8}$ (proposed by Halenke and Möslinger), which may be conveniently written as $T = \frac{D}{4} + \frac{5}{4}F$ (for ease in committing to memory). These last two formulæ do not employ a constant depending on the specific gravity of the total solids not fat, as do the other formulæ given above.

All these formulæ, which however do not apply to old milk, give practically the same results, and should be used to check the results obtained in previous determination of the specific gravity, total solids and fat (pp. 177-185), provided these have been carried out with the same sample of milk.

## Nitrogenous Substances.

The nitrogenous substances (proteins) present in milk are casein, lacto-albumin and lacto-globulin (see p. 175).

The casein may be conveniently precipitated by suitable reagents and estimated quantitatively, while the lacto-albumin is estimated, either from the difference between the total nitrogen and that accounted for by the casein, or by precipitation from milk after removal of the casein. The lacto-globulin, being present in such very small amounts, may be neglected in ordinary laboratory practice.

### 183. *Determination of the Total Nitrogen by Kjeldahl's Process.*

About 10 grams of milk are weighed out from a small beaker into a Kjeldahl flask, and 15 to 20 c.c. of concentrated sulphuric acid added. A few grams of potassium sulphate are added to raise the boiling-point of the acid, and a small crystal of copper sulphate is dropped in to hasten the reaction.

The flask is then heated gently over the naked flame until most of the water has evaporated, after which the mixture is allowed to boil until complete oxidation of all carbonaceous particles has taken place, and the liquid is perfectly clear and transparent.

The flask is then cooled, the contents washed into an ammonia distillation apparatus (p. 12), made alkaline with caustic soda, and the ammonia distilled over into a known volume of standard sulphuric acid $\left(50 \text{ c.c. } \frac{N}{10}\right)$ in the usual way. The excess of acid is then titrated, and the percentage of nitrogen in the milk calculated.

As casein contains 15·65 per cent. nitrogen and lacto-albumin 15·77 per cent., the percentage of total protein may be obtained by multiplying the percentage of total nitrogen by 6·38 $\left(= \frac{100}{15 \cdot 66}\right)$.[1]

The total protein determined thus is always subject to a slight error, owing to the presence of small quantities of nitrogenous substances other than proteins in milk.

[1] Total protein = 85 per cent. casein, and 15 per cent. lacto-albumin, hence 85 per cent. of 15·65 + 15 per cent. of 15·77 = 15·66.

## Estimation of Casein.

Casein (or caseinogen) is present in milk, not in ordinary solution but probably in a state of colloidal suspension, since it can be completely removed from milk by filtering the latter through unglazed porcelain. It is combined with calcium salts, and these calcium casein compounds are precipitated by the action of rennet on milk. A somewhat similar action takes place on saturating milk with various metallic salts, such as $MgSO_4$, NaCl, etc. Coagulation of milk with acids, however, results in the throwing down of casein free from lime salts.

Casein possesses acidic properties, in that it dissolves in dilute solutions of alkalies, and replaces carbon dioxide from metallic carbonates.

To 10 grams of milk are added, with constant stirring, about 10 c.c. of a saturated solution of $MgSO_4$ in water, and solid $MgSO_4$ is then added until saturation of the whole volume of liquid is attained.

The precipitated casein is filtered from the whey, and is washed several times with saturated $MgSO_4$ solution. The solid is then washed into a Kjeldahl flask, and the percentage of nitrogen estimated as usual, by heating with concentrated sulphuric acid, and distilling after rendering alkaline.

The proportion of nitrogen in casein being 15·65 per cent., the factor 6·38 $\left(= \frac{100}{15 \cdot 65}\right)$ is employed for converting the percentage of nitrogen into percentage of casein.

## Analysis of Sour Milks.

When milk is kept for some time, a certain quantity of lactic acid is produced by fermentation of the milk sugar, and the milk becomes acid or "sour." The lactic acid causes a coagulation of the casein, and the ordinary methods of analysis cannot be conveniently applied directly to the coagulated sample.

A good method of dealing with milk in this condition is that proposed by Weibull.

184. *Weibull's Method.*

The approximate volume of the sample is estimated by inspection and 10 per cent. of this volume of ammonia solution is added. On shaking, the casein will dissolve, and the liquid becomes homogeneous.

The volume of the ammoniacal milk is then accurately measured by means of a measuring glass or burette, and by subtraction of the volume of ammonia solution added the original volume of the milk is obtained.

The ordinary analytical operations, such as determinations of total solids, fat, casein, etc., may be carried out with the mixture of milk and ammonia, and from a knowledge of the percentage of the original milk in the ammoniacal liquid accurate values can be obtained for the milk.

In the case of the determination of the density, allowance must be made for the density of the ammonia solution added.

This is best done by means of the formula originally proposed by Weibull :—

$$D=\frac{(M+A)D'-AD''}{M}$$

where

$D$ = density of original milk
$D'$ = density of ammoniacal mixture
$D''$ = density of ammonia
$M$ = volume of milk
$A$ = volume of ammonia.

## Adulteration of Milk.

As distinct from the addition of those substances employed as preservatives, the commoner methods of adulteration or falsification are as follows :—

(i.) The addition of water.

(ii.) The removal of cream.

(iii.) The addition of calcium saccharate (employed to increase the viscosity, and hence give the appearance of richness in cream).

(iv.) The addition of artificial colouring matter.

The first two methods of falsification, being those of which the detection is most important, will alone be dealt with here.

The density of pure fresh milk being on the average 1·032, while the density of the fat globules is about ·86, it follows that the removal of cream will increase the density, while the addition of water will decrease it. It is obvious, therefore, that by the simultaneous removal of cream and addition of water in correct proportion, the density will remain unchanged, the increase by the removal of fat being exactly balanced by the decrease following the addition of water.

The determination of the density alone, therefore, will not lead to the detection of any such adulteration.

### **185.** *Change in Density of Milk on Skimming and Diluting.*

About 200 c.c. of fresh milk are taken, and a small proportion of cream thoroughly stirred in, in order to render more apparent the effect of skimming on density.

The milk is then divided into two equal portions, one of which is allowed to stand for two to three hours in order to permit of the rising to the surface of the cream. The density of the other portion is determined, preferably by means of Westphal's balance (**177**, p. 179).

After two to three hours, as much of the cream as possible is removed from the surface of the first portion by means of a pipette, the skim milk remaining thoroughly mixed by pouring from one vessel to another, and its density determined as before. It will be found that the density has increased, owing to the removal of the lighter fat globules.

To 50 c.c. of the skim milk is then added 1 c.c. (2 per cent.) of water, the whole thoroughly mixed, and the density again determined. Another 1 c.c. of water is then added as before, and the process repeated three or four times. It will be found that the density regularly decreases as the proportion of water is increased, and the results obtained should be plotted in the form of a curve showing the relation between percentage of added water and density. From the curve can be obtained the percentage of water necessary to give a density for diluted skimmed milk of the same value as that possessed by the original whole milk before skimming.

*Detection of Added Water.*—The addition of water to milk will lead to a decrease in the density, the fat contents and the total solids, but the proportion of fat in the total solids will not have undergone any alteration.

Milk should generally be suspected of having been adulterated by the addition of water, if at the same time—

| | | |
|---|---|---|
| the density is below . . | 1·028 | |
| the total solids are below . | 10·3 | per cent. |

*Detection of Removal of Cream.*—The removal of cream from milk or the addition of skim milk will generally lead to a certain increase in the specific gravity, but will always have a marked effect on the percentage of fat, and especially on the proportion of fat in the total solids. The solids not fat will not generally have been decreased.

Milk may come under suspicion of having been skimmed, or diluted with skim milk, if the percentage of fat in the total solids is below 19·9, or—

| | | |
|---|---|---|
| the fat is below . . . | 2·5 | per cent. |
| the solids not fat are above . | 8·0 | " |

*Detection of Simultaneous Removal of Cream and*

*Dilution* (with water or skimmed milk).—As shown above, the density may have increased, decreased, or undergone no change, depending upon the amount of cream removed and the proportion of water added.

The removal of cream will have led to a decrease in the percentage of fat. Dilution with water would also lower the fat content, but would not lead to any change in the proportion of fat in the total solids, other than that accounted for by the removal of cream; dilution with skim milk would, however, show a further decrease in the percentage of fat in the total solids.

### 186. Preservatives.

The addition of any preservatives whatever to milk is illegal. The following tests for four of the commoner preservatives should be carried out with four samples of milk, to which the following preservatives have been added in the proportions given :—

No. I. 5 milligrams of boric acid per 10 c.c. milk.
II. 5 drops of formalin per 100 c.c. milk.
III. 5 drops of hydrogen peroxide per 100 c.c. milk.
IV. .1 gram salicylic acid per 100 c.c. milk.

*Detection of Boric Acid or Borates.*—The first sample of milk (100 c.c.) is evaporated to dryness, and gently ignited. The ash is then dissolved in a small quantity of dilute HCl, filtered free from carbon particles, and evaporated completely to dryness.

(i.) A portion of the residue is dissolved in a little very dilute HCl, and a piece of turmeric paper dipped into the solution. The presence of boric acid is shown by the fact that the paper will turn reddish brown on drying in a steam oven, and this coloration can be distinguished from that produced by the alkalies from the fact that when moistened with a drop of an alkaline solution, the brown colour is changed to greenish black.

(ii.) A portion of the residue is placed in a test-tube with a small quantity of concentrated $H_2SO_4$, a little alcohol added, and the mixture boiled. On igniting the alcohol vapour, a green colour will be noticed indicative of the formation of a volatile boron compound.

*Detection of Formaldehyde.* — One drop of dilute $HNO_3$ is added to 10 c.c. of the milk from sample II. above, and a few c.c. of concentrated $H_2SO_4$ poured down the side of the test-tube.

A violet-blue ring will be found at the junction of the two liquids. This colour reaction between formaldehyde and nitric acid in the presence of sulphuric acid does not take place except in the presence of proteins, and is therefore not given in aqueous solutions.

Schiff's reagent (rosaniline decolorised by $SO_2$) may also be used for the detection of formaldehyde in milk. The milk is curdled with dilute $H_2SO_4$ and the filtrate added to Schiff's reagent. The formation of a pink colour shows presence of formaldehyde.

*Detection of Hydrogen Peroxide.*—Ten c.c. of the milk that has been treated as in III. above are placed in a test-tube, and a few drops of an alcoholic solution of benzidine added. A blue coloration shows the presence of $H_2O_2$.

*Detection of Salicylic Acid.*—A few c.c. of the milk are treated with sufficient mercuric nitrate solution to precipitate all the casein, and the mixture filtered. The filtrate is then shaken up with ether in a large test-tube, the ethereal solution poured off, divided into two parts, and the ether allowed to evaporate. The residue is treated in the one case with one drop of neutral dilute ferric chloride, when a violet coloration shows presence of salicylic acid. To the other residue is added a little bromine water, when a yellowish precipitate of tribromo-salicylic acid confirms the presence of salicylic acid.

# CHAPTER XX

## BUTTER

BUTTER consists of the fat globules of milk which have been caused to coalesce and revert from their super-cooled and emulsified condition to the solid form by the action of churning or other mechanical treatment. As a result of churning, the butter is obtained in the form of granules about the size of small shot, but by subsequent processes of treatment it is worked up into a homogeneous mass.

In addition to the pure fat, butter usually contains a certain percentage of water, and also a small quantity of other milk constituents, such as protein and milk sugar.

The relative proportions of the constituents vary slightly, and depend upon whether the butter has been prepared from fresh or ripened cream, while the percentage of ash depends largely upon the amount of common salt added.

*Average Composition of Butter from Fresh Cream (Unsalted).*

| | |
|---|---|
| Fat | 85·0 |
| Water | 13·92 |
| Protein | ·60 |
| Milk sugar | ·35 |
| Ash (NaCl, etc.) | ·13 |

### 187. *Determination of Water.*

From 15 to 20 grams of recently ignited pumice in small pieces are placed in a porcelain dish, which is then heated in the steam oven for an hour, and weighed together with its contents after cooling in a desiccator.

About 10 grams of butter are placed in a stoppered bottle, and melted by immersing the latter in water at about 40° C. The liquefied butter is then shaken violently, while allowed to cool until it solidifies, and about 5 grams of the solid are then placed on the pumice, and the dish and its contents again weighed. The weight of butter actually taken is thus determined.

The whole is heated in the steam oven for an hour, cooled in the desiccator, weighed, then reheated in the oven for another half-hour, and the process repeated until no further loss in weight occurs.

The combined duration of heating should not be too long, as part of the butter may undergo oxidation, resulting in an actual increase in weight from the minimum.

### 188. *Determination of Fat.*

The material left in the basin after the estimation of water according to the above method may be conveniently employed for the determination of the fat, provided too great a time be not allowed to elapse.

The pumice is transferred into a previously extracted filter-paper cartridge, and the cartridge and contents placed in a Soxhlet extractor in the bottom of which is placed a layer of extracted cotton-wool.

To ensure complete removal of the fat from the basin, the latter should be washed out several times with ether, which is then poured into the cartridge while the latter is in the extractor. Another layer of extracted cotton-wool is placed on the top of the cartridge, and the whole extracted with dry ether on the water-bath in the usual way, a weighed flask having been fitted to the extractor. After a few hours the flask and its contents are removed, the ether entirely evaporated off in the water-bath, the flask placed for a few minutes in the steam oven, then cooled in the desiccator, and weighed. The increase in weight of the flask will give the amount of fat.

The extracted residue left in the cartridge after the above operation may, if necessary, be used for the determination of common salt (see **190** below).

**189.** *Determination of Total Solids not Fat.*

The solids other than fat present in butter include casein, milk sugar, lactic acid, and ash constituents (NaCl, etc.), and the estimation of the percentage is of importance in the detection of adulteration by certain methods that increase the total solids other than fat. For example, adulteration with milk powder would increase the percentage of milk sugar, etc.

From 10 to 15 grams of butter are weighed out into a porcelain basin, and melted at as low a temperature as possible.

A few c.c. of dry ether is poured on to the butter, the mixture thoroughly stirred, and the ethereal solution of fat filtered through a filter paper that has been previously extracted with ether, dried at 100°, and weighed.

As much as possible of the residue insoluble in ether should be retained in the basin, and treated successively with further quantities of ether, which are poured as before on to the filter paper. Finally, the contents of the basin are well washed with ether into the filter, which is then carefully folded up, into the form of a cartridge, and extracted with ether in a Soxhlet extractor for about half an hour, in order to remove all traces of fat. The filter paper and contents are then removed from the extractor, dried in the steam oven, and weighed. The increase in weight will give the amount of solids other than fat present in the quantity of butter taken.

**190.** *Determination of Common Salt.*

The filter containing casein, NaCl, milk sugar, etc., obtained after the determination of total solids other than fat is fitted into a glass funnel of suitable size, and the contents well washed with boiling water from a wash-bottle into a small conical flask.

The aqueous solution so obtained will contain milk

sugar, NaCl, and other matter soluble in water, and the amount of common salt is then estimated in the solution by direct titration with $\frac{N}{10}$ silver-nitrate solution, using a few drops of potassium chromate as indicator. One c.c. $\frac{N}{10}$ silver nitrate is equivalent to ·00585 gram NaCl.

(The approximate percentage of casein in the butter may be obtained by drying and weighing the filter paper after washing with water, and subtracting from this weight the known weight of the filter paper.)

**191.** *Determination of Protein.*

The total protein in butter may be best determined by the ordinary Kjeldahl process after removal of the majority of the fat, as this substance is liable to cause excessive frothing.

About 20 grams of butter in a porcelain basin should be treated with ether according to the method described above (see **189**, p. 198) for the estimation of total solids not fat, and the residue washed on to a filter paper as before.

The filter and contents are then heated in a Kjeldahl flask with 15 c.c. of concentrated $H_2SO_4$, and the operation completed as usual (see p. 10). The percentage of nitrogen multiplied by 6·38 will give the percentage of protein in the butter.

### Reichert-Meissl Number.

Butter-fat is a mixture of the glycerides of a number of different fatty acids (see p. 184), and differs from other fats, such as those which might be employed to adulterate it, by reason of the fact that a considerable proportion of the fatty acids combined with the glycerine are of low molecular weight, *and readily volatile in steam.* These facts are made use of for the detection of adulteration with other fats, the presence of which will generally tend to lower the proportion of the glycerides of the volatile fatty acids present.

This is the basis of the process originally proposed by Reichert, and modified later by Meissl, Wollny, and Leffmann and Beam.

The Reichert-Meissl number may be defined as the number of c.c. of $\frac{N}{10}$ alkali solution equivalent to the fatty acids volatile in steam and present in 5 grams of butter.

**192.** *The Determination of the Reichert-Meissl Number by the Modified Method of Leffmann and Beam.*

Exactly 5 grams of butter-fat from a sample which has been melted and filtered free from casein, etc., are weighed out into a 300 c.c. round-bottomed flask, and 20 c.c. glycerine and 2 c.c. caustic-soda solution, containing 1 gram NaOH per c.c., are added.

By gently heating the flask the water is driven off, as shown by the fact that the solution becomes clear, and the heating should be continued until this clarification has taken place. The glycerides present in the butter have now been hydrolysed to glycerine and the sodium salts of the various fatty acids. The fatty acids themselves are then liberated from the sodium salts by dissolving the latter in 90 c.c. hot water, and adding 50 c.c. of dilute $H_2SO_4$ ($2\frac{1}{2}$ per cent. concentrated acid by volume) to the mixture in the flask. The flask is then connected to a water-trap fitted to a condenser (see Fig. 28), and distillation immediately commenced into a flask graduated on the neck to 100 c.c. and 110 c.c.

The rate of distillation should be so adjusted that the 110 c.c. mark is reached in from thirty to forty minutes. The distillate is then filtered through a dry filter, and 100 c.c. titrated with $\frac{N}{10}$ caustic-soda solution, using phenolphthalein as indicator.

By adding one-tenth to the number of c.c. of caustic soda employed the amount that would have been required for the whole distillate is obtained. This value is known as the Reichert-Meissl number.

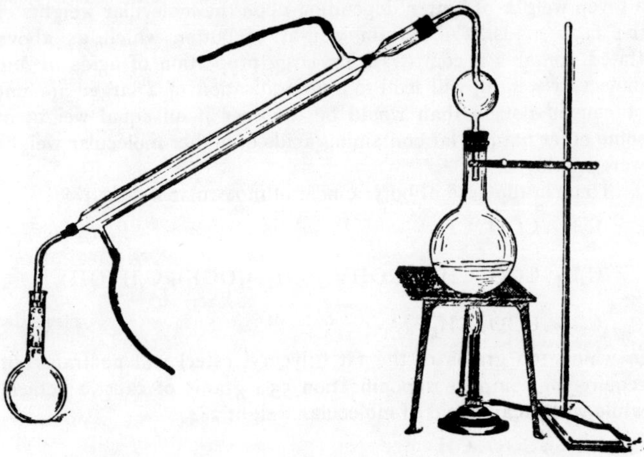

FIG. 28.—Apparatus for determining the Reichert-Meissl value of fats.

A complete determination of the Reichert-Meissl number, as described above, should also be carried out, for the sake of comparison, with 5 grams of margarine.

The time of distillation, the volume of liquid collected, etc., are purely arbitrary figures, but the directions should be exactly followed out in order that comparable numbers may be obtained. In the original Reichert-Meissl process, alcoholic potash was employed for the hydrolysis in place of aqueous soda and glycerine.

The Reichert-Meissl number for pure butter-fat is generally between 24 and 34, but may in exceptional cases fall as low as 19.

In the case of most animal fats, however, it is seldom above 1, and adulteration of butter with margarine or other fats may thus be readily detected in this manner.

### Saponification of Butter-fat.

When butter-fat is boiled with caustic alkalies, saponification takes place, with production of glycerol and the sodium salts of the fatty acids (sodium soaps). A certain proportion of the caustic alkali is thus neutralised, the amount undergoing neutralisation by a given weight of butter depending upon the molecular weights of the fatty acids. A certain weight of butter, which, as above stated, contains a comparatively large proportion of acids of *low* molecular weight, will lead to a neutralisation of a larger amount of caustic potash than would be the case if an equal weight of some other natural fat containing acids of higher molecular weight were employed.

Thus, in the case of butyric acid of molecular weight 188,

$$\begin{array}{l} C_3H_7.COO.CH_2 \\ \qquad\qquad\quad | \\ C_3H_7.COO.CH + 3KOH = 3C_3H_7.COOK + C_3H_5(OH)_3 \\ \qquad\qquad\quad | \\ C_3H_7.COO.CH_2. \end{array}$$

in which 100 grams of the fat (glyceryl ester) will neutralise or require for complete saponification 53·4 grams of caustic potash, while with stearic acid of molecular weight 284,

$$\begin{array}{l} C_{17}H_{35}.COO.CH \\ \qquad\qquad\quad | \\ C_{17}H_{35}.COO.CH + 3KOH = 3C_{17}H_{35}.COOK + C_3H_5(OH)_3 \\ \qquad\qquad\quad | \\ C_{17}H_{35}.COO.CH_2. \end{array}$$

in which 100 grams of the fat will require 18·8 grams of caustic potash. Use is made of this principle in a method proposed by Koettstorfer for the quantitative examination of butter, it being obvious from the above equations that adulteration of butter by means of other natural fats (of higher average molecular weight) leads to a smaller amount of caustic potash being required for saponification per cent. of fat. The results are usually expressed as the Koettstorfer number, this being the number of milligrams of caustic potash required per 1 gram of butter-fat. In the case of glyceryl-tributyrate the Koettstorfer number would be 534, while with glyceryl-tristearate the value would be 188. For pure butter-fat the Koettstorfer number generally lies between 220 and 233, while other fats (*e.g.* margarine) and oils have lower values, in the neighbourhood of 190 to 200.

The Koettstorfer number is sometimes known as the "saponification number," the term "saponification equivalent" being employed to express the mean molecular weight of the fatty acids.

### **193.** *The Determination of the Koettstorfer or Saponification Number.*

In this determination it is advisable that a blank experiment be performed at the same time as the actual experiment, using similar apparatus.

Two round-bottomed 150 c.c. flasks are fitted with corks, each provided with a glass tube about 3 feet in length, to act as an air condenser. An alcoholic solution of caustic potash is prepared by dissolving about 30 grams of caustic potash in 1 litre of absolute alcohol.

Two grams of butter-fat are then weighed out into one of the flasks, and thereafter the two flasks treated in exactly similar ways as follows:—

Twenty-five c.c. of the alcoholic caustic potash are run into each flask from a burette, the air condensers fitted, and the solutions boiled gently for a quarter of an hour.

When the boiling has proceeded for the time mentioned, the heating is discontinued, the flasks disconnected from the condensers, and their contents titrated with the $\frac{N}{2}$ hydrochloric acid, using phenolphthalein as before. The number of c.c. of acid required by the contents of the flask to which the butter was added are subtracted from the number of c.c. required by the blank flask, and the amount of alkali required for the saponification then calculated. The weight in milligrams of caustic potash required to saponify 1 gram of butter is then given as the Koettstorfer number of the sample of butter employed.

**194.** *The Unsaturated Acids in Butter-fat.*

While most of the acids present as glycerides in butter are saturated and contain no double or triple bonds, there are, however, always present a certain small proportion of unsaturated acids, notably oleic, and for this reason butter-fat will show a certain absorption for iodine, generally to the extent of 25 to 50 per cent. It is usual in butter examination to determine the iodine value, as valuable indication may be thus afforded as to whether the butter has been adulterated with any fats or oils which show a higher iodine absorption.

Hübl's method, described in **13**, p. 22, may be employed on 0·5 gram of butter, following out the details given; but Wijs' method is more modern, and possesses certain notable advantages over the older method with regard to constancy of results and quickness of determination.

**195.** *Determination of the Iodine Value by Wijs' Method.*

In this process the iodine is employed in the form of iodine chloride (ICl).

*Iodine Chloride Reagent.*—9·4 grams of iodine trichloride are weighed into a flask of about 300 c.c. capacity, into which is then poured about 200 c.c. of glacial acetic acid; the flask, fitted with a cork through which passes a calcium chloride tube, is heated on the water-bath until all is dissolved. 7·2 grams of finely powdered iodine are dissolved in glacial acetic acid in a similar manner and added to the first solution. The mixture is stoppered, allowed to cool, and made up to 1 litre with acetic acid. It is titrated after twenty-four hours as follows :—Ten or 20 c.c. of solution are mixed with 5 to 10 c.c. of potassium-iodide solution, diluted with 300 c.c. or so of distilled water, and titrated with $\frac{N}{10}$ sodium-thiosulphate solution in the ordinary manner. The strength of the iodine-chloride solution alters during the first twenty-four hours, but after that will remain constant for several weeks. It must be restandardised from time to time.

*Iodine Value Estimation.*—About 0·75 gram of butter-fat is accurately weighed out and dissolved in 10 c.c. of carbon tetrachloride, and when dissolved 25 c.c. of the Wijs' solution is added and the flask stoppered. After standing in the dark for one to two hours the contents are washed into a conical flask with water and mixed with 10 c.c. of KI solution, the total volume being about 300 c.c. The residual iodine is titrated with standard sodium thiosulphate.

The calculation is made as previously described for the Hübl method (**13**, p. 23).

## The Colour of Butter.

Butter possesses naturally a light yellow colour, the deepness of which depends on several causes. It is generally found that, during the summer months when the cows are on grass, the colour is deeper, while during the winter when the diet consists largely of concentrated foods, together with hay, straw, and roots, butter may be practically colourless. Again, the natural colour of butter depends upon the breed of cow, the butter produced from the milk of the Channel Islands breeds (Jersey, Guernsey) being notably deeper in colour. It is the custom to colour butter artificially, in order that it may be put on the market with the same colour all the year round, and the artificial colouring matter used for this purpose is most usually annatto, a natural yellow substance obtained from the fruit-capsules of *Bixa orellana.* The dye is dissolved in olive or other oil, and the solution added in small amount to the cream before churning. Other natural colouring matters are sometimes employed and may be regarded as innocuous; the use of aniline dyes is, however, to be deprecated, while lead chromate, which has been used, is distinctly poisonous.

**196.** *Detection of Artificial Colouring Matter.*

*Annatto.*—About 5 grams of butter are dissolved in 50 c.c. ether, and about 10 c.c. of dilute caustic soda added. The mixture is thoroughly shaken, allowed to stand for a short time, and the brown alkaline liquid containing the colouring matter dissolved in it is separated off and evaporated to dryness. One drop of strong sulphuric acid is then added, and the production of a blue colour changing to green shows the presence of annatto.

*Yellow Aniline Dyes.*—A few grams of butter are melted and shaken up with an equal volume of strong HCl. The acid will be coloured red in the presence of most aniline dyes.

# CHAPTER XXI

## CHEESE

THE casein in milk appears to exist in a state of colloidal suspension, and is loosely combined with the calcium salts also present. These calcium compounds undergo a change under the action either of rennet or acids, whereby there is precipitated an insoluble curdy substance, the nature of which varies in the two cases.

In the case of the precipitation by acids, the latter form soluble calcium salts with the lime present in the casein complex, and casein is thrown down in a more or less pure condition. The action of rennet is not so simple, for, according to Hammersten, rennet converts casein into two cleavage products, paracasein and whey albumin, the former in largest amount.

In the absence of soluble calcium salts the paracasein is not precipitated, but as these salts are invariably present in milk, a curd is produced which is essentially different in character to that produced by mineral acids or by the lactic acid formed by fermentation during the souring of milk.

The curd produced by rennet may be regarded as crude cheese, which in cheese-making is submitted to cutting, pressing, and other mechanical treatment, and is finally allowed to ripen. The chemical composition of cheese varies largely, depending upon many factors, such as nature of milk employed (whole milk or skim), variations in pressing, salting, ripening, etc.

### 197. *Estimation of Water.*

About 20 grams of washed and ignited sand or asbestos are placed in a porcelain basin, heated in the steam oven for half an hour, cooled in the desiccator, and weighed. Three to 5 grams of cheese in small pieces are then weighed out into the basin, and the whole heated for three hours in the steam oven. The dish and its contents are cooled in a desiccator and weighed, and the process repeated exactly as described in the case of milk or butter (pp. 183 and 196). If the heating be carried out for too long a period, there is danger of loss of volatile fatty acids. A small loss of ammonia invariably occurs, but is negligible.

The percentage of water in cheese naturally varies largely, but is generally between 26 and 46.

*Approximate Percentages of Water in Different Cheeses.*

| | |
|---|---|
| Stilton | 30 |
| Gruyère | 35 |
| Dutch | 37 |

198. *Estimation of Fat.*

The material remaining in the basin after the estimation of moisture (above), is transferred by means of a spatula to a fat-free cartridge, and the latter placed in a Soxhlet extractor. The basin is well washed with ether, the washings being poured on to the cartridge in the extractor.

Extraction with ether is then carried out as usual for two hours, the contents of the cartridge then placed in a mortar, well ground, returned to the cartridge, the mortar washed out as before, and the extraction proceeded with. This procedure is necessary as a certain proportion of the fat globules are embedded in the mass of the cheese, and hence do not come into contact with the ether. It may be necessary to repeat this grinding three or four times, until it is found that the contents of the weighed flask, after driving off the ether, do not increase further in weight.

A modification of Gerber's method for the estimation of fat in milk (p. 185) may also be employed in the case of cheese.

*Approximate Percentages of Fat in Different Cheeses.*

| | |
|---|---|
| Cream | 40 |
| Gruyère | 30 |
| Dutch | 20 |
| Skim-milk | 10 |

**Nitrogenous Substances.**

The nitrogenous substances present in cheese may be divided as follows: (i.) insoluble proteins; (ii.) water-soluble proteins, amides, etc.; (iii.) ammonia and ammonium compounds.

During the process of ripening of cheese the proportion of these different nitrogenous constituents varies, the general tendency being to split up complex insoluble proteins into simpler soluble proteins or amino-bodies, and ammonia. These changes are due to the action of micro-organisms and also of enzymes, which show a greater activity in rennet cheeses than in the case of cheeses produced from the curd precipitated by acids. Cheeses of the latter type are practically unknown in this country.

**199.** *Determination of Total Nitrogen.*

The percentage of total nitrogen present in cheese is determined by Kjeldahl's process (p. 10), using 3 to 4 grams of cheese.

**200.** *Determination of the Total Soluble Nitrogenous Compounds.*

Ten grams of cheese are rubbed to a paste with warm water, and the liquid transferred to a 250 c.c. standard flask, which is then filled up to the mark. The flask and its contents are allowed to stand for some hours, and the liquid is then filtered through a dry filter. Fifty c.c. of the clear filtrate are withdrawn by means of a pipette, transferred to a Kjeldahl flask, evaporated to dryness, 20 c.c. concentrated sulphuric added, and the nitrogen estimated by Kjeldahl's process. The soluble proteins are estimated in another portion of the original filtrate as follows :—

**201.** *Determination of Soluble Protein Nitrogen.*

Fifty c.c. of the clear filtrate containing the water-soluble constituents of cheese are acidified with dilute $H_2SO_4$, and 20 c.c. of a 10 per cent. solution of phosphotungstic acid added.

The precipitated proteins are filtered off, well washed with water, and the nitrogen estimated by Kjeldahl's process, the result being given as nitrogen in the form of water-soluble proteins.

**202.** *Determination of Ammoniacal Nitrogen.*

One hundred c.c. of the clear filtrate containing the water-soluble constituents of cheese are placed in an ammonia distillation apparatus (p. 13), and excess of magnesia added.

The ammonia is then distilled into a known volume of standard acid, and the amount of acid neutralised estimated by titration as usual.

Magnesia is employed in the place of potash or soda, as these latter substances cause a hydrolysis to ammonia of some of the amino-bodies present; magnesia does not act in this way.

### 203. *Determination of Nitrogen in Form of Amides.*

The value for the nitrogen in the form of amides is obtained by subtracting from the total water-soluble nitrogen (**200** above) the soluble protein (**201**) and the ammoniacal nitrogen (**202**).

N % in form of amides = total water-soluble nitrogen % *minus* N% in form of soluble protein *minus* N% in form of ammonia and ammonium compounds.

### 204. *Determination of Casein.*

The percentage of casein may be determined by taking the difference between the percentages of total nitrogen and water-soluble nitrogen and multiplying the result by 6·38.

According to the above method of analysis, the nitrogen in cheese may be differentiated as follows:—

Total nitrogen.
Nitrogen in form of casein.
Nitrogen in form of water-soluble proteins.
Nitrogen in form of amides.
Nitrogen in form of ammonia and ammonium compounds.

The above differentiation between the several forms of combined nitrogen should, for the sake of comparison, be carried out with an immature or "green" cheese and also with an over-ripe sample.

### 205. *Determination of Ash.*

About 15 grams of cheese are carefully ignited in a platinum or porcelain basin of known weight in a muffle furnace until all the carbon has been burnt off. After cooling in a desiccator the basin and contents are weighed, and the percentage of ash constituents thus determined.

### 206. *Estimation of Sodium Chloride.*

The common salt is best determined in the ash obtained above, by extracting the latter several times with small quantities of hot water, and filtering the aqueous solution obtained into a 100 c.c. flask, and filling up to the mark. Portions of 25 c.c. are withdrawn, and titrated as usual with $\frac{N}{10}$ $AgNO_3$ solution, using potassium chromate as indicator.

### 207. *Detection of Phosphoric Acid in Ash Constituents.*

The insoluble residue remaining on the filter after the washing is examined for phosphoric acid as follows:—

A portion of the residue is dissolved in fairly concentrated $HNO_3$ in a test-tube, ammonium molybdate added, and the solution warmed. The characteristic canary-yellow precipitate of ammonium phosphomolybdate shows the presence of phosphates in the ash, these phosphates being derived from phosphoproteins, etc., in the cheese.

## Rennet.

Rennet, which is largely employed for the manufacture of cheese owing to its action in precipitating casein (paracasein), owes its activity in this respect to an enzyme, or unorganised ferment, named *rennin* or *chymosin*, which occurs in the stomachs of young animals living on milk. Rennet is generally prepared from the fourth stomach of the calf by extracting the dried stomach with dilute salt solution. Rennet is largely employed directly in the form of this solution, or may be purchased as more concentrated and purer solutions, or in the solid form as pellets or powder, which has been precipitated from the saline solution by addition of further quantities of salt.

The rapidity of precipitation of casein (or more correctly paracasein, see p. 207) by rennet depends upon several factors, such as the "strength" of the rennet, the temperature, and the degree of acidity (p. 182) of the milk. The more acid the milk, the more rapid the action, while the optimum temperature is about 40° C. A temperature as high as this, however, is never employed in practice, as it would be almost impossible to retain the fat in the curd. A temperature of 29° C. (84° F.) is most commonly used.

### 208. *Determination of Rennet Strength.*

Rennet strength may be defined as the volume in c.c. of milk which will be coagulated in forty minutes, at a temperature of 35° C., by 1 c.c. of rennet solution or 1 gram of rennet powder.

Five c.c. of rennet solution, or 5 grams of rennet powder, are made up to 100 c.c. with distilled water, and 1 c.c. of this solution is withdrawn by means of a pipette, and forcibly blown into 100 c.c. of separated milk warmed to 35° C., and contained in a beaker, which is immersed in a larger beaker containing water at the same temperature. The mixture is well stirred by means of a thermometer, and the exact time[1]

[1] The formation of a coagulum may be accurately noted by placing a piece of straw or splinter of wood on the surface, when the stirring is finished. The liquid will continue to rotate until coagulation occurs, when the splinter will be seen suddenly to come to rest.

(in seconds) noted between the addition of the rennet and the commencement of the curdling. The strength of the rennet is then calculated by simple proportion, as shown by the following example.

Time between addition of rennet and coagulation $=6$ minutes, or 360 seconds.

1 c.c. rennet solution contains $\frac{5}{100} = \cdot05$ grams (or c.c.) of rennet.

Now ·05 gram rennet in 360 seconds coagulate 100 c.c. milk,

therefore 1 gram rennet in $40 \times 60$ seconds coagulate

$$\frac{100 \times 1 \times 40 \times 60}{\cdot05 \times 360} = 13{,}333\cdot$$

The rennet strength in this case is, therefore, 13,333·

Rennet solutions may possess strengths up to 10,000, while rennet powder may attain a value of 300,000. The reaction being a catalytic one, a small proportion of rennet may theoretically coagulate an infinite volume of milk, although the time taken for coagulation will vary inversely with the amount of rennet employed.

# SECTION VI.—EXAMINATION OF WATERS AND SOAP

## CHAPTER XXII

### Analysis of Water

**The** complete analysis of water includes the determination of :—

(1) Suspended matter.
(2) Total dissolved solids.
(3) Organic matter, free and albuminoid ammonia.
(4) Nitrates and nitrites.
(5) Chlorides.
(6) Hardness, temporary and permanent.

It is also sometimes necessary to determine the oxygen absorbed by the organic matter present and make a quantitative examination of the dissolved solids. According as the water is to be used for potable or industrial purposes, or for the preparation of sprays, certain of these estimations may be left out.

A fair average sample of the water must be taken, and for a complete analysis a couple of Winchesters should be filled, stoppered, and kept in a cool dark place until required. If the hardness only is to be determined, a much smaller sample will suffice. The analysis should follow as soon as possible after the collection of the sample.

#### Preliminary Examination.

**209.** *Suspended Matter.*

If the water contains much suspended matter, this is estimated by filtering 250 c.c. of the well-shaken sample through a washed tared filter paper, or a Gooch crucible. The residue is washed with distilled water, dried, and weighed.

Previous to drying, valuable information may be gained as to the character of the suspended matter by examination under the microscope.

*Colour.*—The colour or turbidity of the filtered water should be examined by looking down a 2-foot tube, placed side by side with a similar tube containing distilled water. A tintometer may be used if available.

The measure of the colour is sometimes a very useful indication of the purity of a water. Yellow or brown tints generally indicate sewage pollution (except in water from peat), or presence of much iron. Green is sometimes formed by organisms containing chlorophyll.

*Taste and Smell.*—Smell is observed after gently warming the water in a closed flask on the water-bath to 40° to 50° C. The taste should be observed at 10° to 20° C. and compared with a good tap-water, not distilled water.

### Total Dissolved Solids.

**210.** Fifty to 250 c.c. of the filtered water are evaporated to dryness in a weighed nickel or platinum dish on the water-bath. The residue is dried in the steam-oven at 100° C., cooled in a desiccator, and weighed. The volume of water taken depends, of course, on the amount of dissolved solids estimated to be present. The total solids residue should be carefully ignited at a dull red heat. If much organic matter is present, blackening will be observed and considerable loss of weight will occur. Too high a temperature must not be employed in estimating this loss, or some of the carbonates will be decomposed and the chlorides volatilised. The loss on ignition should not exceed 20 per cent. of the total solids, but the information obtained from the test is not particularly reliable.

In a good water for general purposes the dissolved matter will not exceed 50 parts per 100,000. Higher values than this indicate excessive hardness (see **222**, p. 230).

### Organic Matter.

**211.** The organic matter present in water is generally estimated by measuring the free or saline ammonia, the albuminoid ammonia, and the amount of oxygen absorbed in a given time. From these three figures a general idea may be obtained of the condition, nature, and source of the organic matter, as well as a fair idea of its actual amount.

*Reagents.*—For the estimation of the ammonia, free and combined, the following reagents are required.

(*a*) *Nessler Reagent.*—This is a solution in potassium hydroxide of a double salt of mercuric and potassium iodides, which gives with ammonia a reddish brown precipitate or a coloured solution, according to the concentration. The depth of colour of the solution is proportional to the amount of ammonia present. The reagent is prepared by dissolving 62·5 grams of potassium iodide in 250 c.c. of distilled water, and then gradually adding (with constant stirring) a cold saturated solution of mercuric chloride, until a very slight permanent precipitate remains. A cold solution of 150 grams of caustic potash in 150 c.c. of water is then mixed with the above, and the total volume made up to 1 litre with distilled water. The solution is allowed to remain in a closed vessel for about three days, and the clear liquid then decanted into a bottle

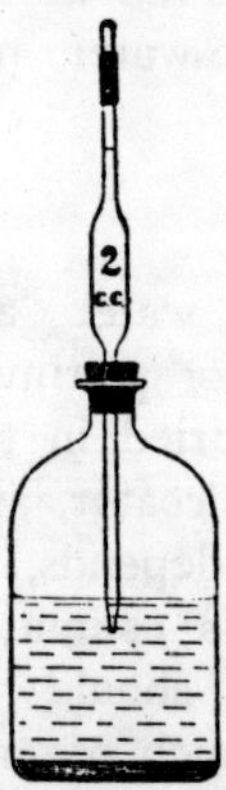


FIG. 29.—Bottle for storing Nessler reagent.

fitted with a rubber stopper through which passes a 2 c.c. pipette (Fig. 29).

(*b*) *Standard Ammonia Solution.*—Two solutions are generally prepared—(1) a stock solution, 1 c.c. of which contains 1 mg. of ammonia, and (2) a solution for use in the tests, 1 c.c. containing 0·01 mg. of ammonia. For the former of these 3·15 grams of freshly sublimed, pure ammonium chloride are dissolved in ammonia-free water, and the solution made up to 1 litre at 15·5° C. The second is prepared by accurately measuring out 10 c.c. of this solution and diluting to 1 litre with ammonia-free water, care being taken that the temperature during the measurements is 15·5° C.

(*c*) *Ammonia-free water* for the tests can be prepared from ordinary tap-water by first adding 10 c.c. of 10 per cent. sodium carbonate solution to a litre of water and then distilling from a flask connected with a Liebig's condenser. The distillate should be rejected until 50 c.c. no longer give a yellow coloration with 2 c.c. of Nessler reagent, after standing for a few minutes. A portion of the remaining water is then distilled into a clean bottle, but the distillation should not be carried too far, as the latter fractions are liable to contain ammonia.

(*d*) *Alkaline Potassium Permanganate Solution.*—This reagent is prepared by dissolving 8 grams of crystallised potassium permanganate and 200 grams of stick potash in 1·5 litres of water. The solution is boiled vigorously in an open dish until about two-thirds of its original volume, and is then allowed to cool and transferred to a bottle.

### *Estimation of Ammonia.*

**212.** *Apparatus.*—The distillation apparatus may be fitted up as shown in Fig. 30. The distilling flask should have a capacity of 1·5 to 2 litres. Rubber stoppers should not be used for closing the flask or for connections. If corks are used it is advisable to cover them with clean tinfoil, and the side-tube from the flask should reach down to the water-cooled portion of the

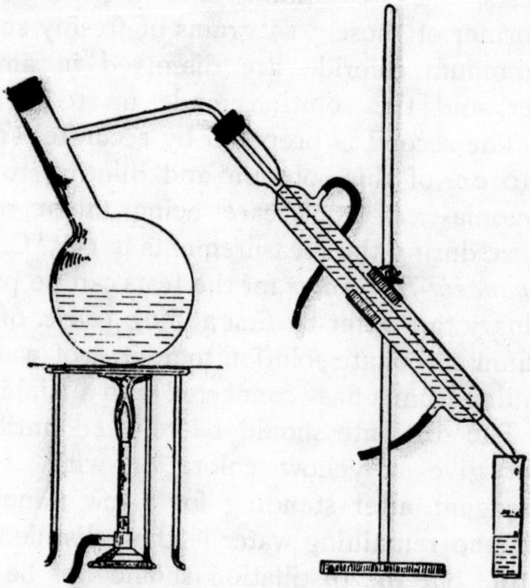

FIG. 30.—Distillation apparatus for water analysis.

inner condenser tube. The distillate is received in Nessler glasses (Fig. 31), uniform cylinders 6 ins. in height and 1½ ins. in diameter, made of perfectly colourless glass, and having a graduation mark at 50 c.c. capacity. Before commencing the estimation the whole apparatus should be thoroughly cleaned by distilling pure water through it, until no trace of any ammonia can be detected in the distillate.

*Free or Saline Ammonia.*—Five hundred c.c. of the water to be examined are measured out and poured into the distilling flask, and about 1 gram of recently ignited sodium carbonate is added. The flask is then closed and the contents boiled. In the first case four Nessler glasses of distillate (each containing 50 c.c.) are collected. The distillation is then stopped and to each of the cylinders is added 2 c.c. of the Nessler reagent, the solutions mixed thoroughly by rotating the glasses, and the latter then placed in the order of collection on a white tile or piece of dull white paper.

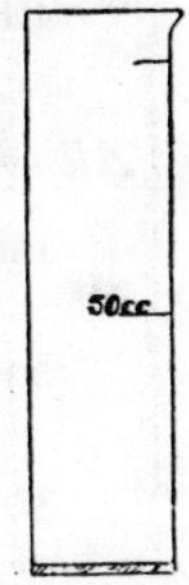

FIG. 31.—Nessler glass.

Into five other Nessler glasses are now run respectively 1, 2, 4, 6, and 8 c.c. of the dilute standard solution of ammonium chloride (=0·01, 0·02, etc., mg. of $NH_3$); the volume is then made up to 50 c.c. in each case with ammonia-free water, and to each is added 2 c.c. of Nessler reagent. After standing for a few minutes the colours of the distillates are compared with those of the standards, and the number of cubic centimetres of ammonium chloride solution required to match the colour of the sample distillate in each case is noted down. The total number of cubic centimetres required is the number of hundredths of a milligram of free or saline ammonia contained in 500 grams of water.

*Albuminoid Ammonia.*—After the above estimation, 50 c.c. of alkaline permanganate solution are added to the water remaining in the distilling flask and the distillation is continued until at least three more Nessler glasses of distillate have been collected. These are treated in the same way as before and compared with the standards.

*Example.*—Five hundred c.c. of water taken for analysis.

*Free Ammonia.*

1st cylinder matched by 2 c.c. standard . . . . = 0·02 mg. $NH_3$
2nd cylinder matched by 0·5 c.c. standard . . . . = 0·005 „
3rd cylinder matched by 0·2 c.c. standard . . . . = 0·002 „
4th cylinder matched by nil . —

0·027 mg. $NH_3$

*Albuminoid Ammonia.*

1st cylinder matched by 4 c.c. standard . . . . = 0·04 mg. $NH_3$
2nd cylinder matched by 2 c.c. standard . . . . = 0·02 „
3rd cylinder matched by 0·5 c.c. standard . . . . = 0·0005 „
4th cylinder matched by nil . —

0·065 mg. $NH_3$

Thus in 1 litre there would be 0·054 mg. free ammonia and 0·130 mg. albuminoid ammonia. As milligrams per litre are parts per million, the water would be stated to contain 0·054 parts per million of free ammonia and 0·130 parts per million of albuminoid ammonia.

### 213. *Interpretation of Results from the Ammonia Estimations.*

The determination of ammonia is important, as both free and albuminoid ammonia nearly always indicate the presence of animal organic matter, generally sewage contamination. Free ammonia is readily oxidised to nitrites and nitrates, and hence its presence, in considerable quantity, usually indicates recent pollution and absence of oxidation.

The albuminoid ammonia is a measure of the nitrogenous organic matter present, since all proteins and many other nitrogenous organic substances yield ammonia when treated with alkaline potassium permanganate. It assumes, of course, that the nitrogenous organic matter in water is uniform in character, and that it conforms to the general rule that "the disintegrating animal refuse may be measured by ten times the albuminoid ammonia which it yields."

In a good drinking water neither free nor albuminoid ammonia should exceed 0·008 part per 100,000.

**214.** *Oxygen Absorbed, or "Moist Combustion."*

The organic matter in water is determined, in terms of the oxygen required to oxidise it, by treatment with potassium permanganate. In contact with oxidisable organic matter in acid solution, permanganate readily gives up its oxygen as follows:—

$$2KMnO_4 + 3H_2SO_4 = K_2SO_4 + 2MnSO_4 + 3H_2O + 5O.$$

The oxidation is not instantaneous, and varies in extent with the temperature.

By means of this method it is possible to distinguish between readily oxidisable first decomposition products and more complex organic bodies, the former being much more rapidly attacked by permanganate at ordinary temperatures than the latter. The solutions required for the estimation are the following:—

1. *Standard Potassium Permanganate* containing 0·396 gram $KMnO_4$ per litre.

   One c.c. of this solution contains 0·0001 gram of available oxygen.
2. *Dilute Sulphuric Acid.*—Strength 1 to 3.
3. *Sodium Thiosulphate* solution containing 1 gram $Na_2S_2O_3$ per litre.
4. *Potassium Iodide solution*, 10 per cent strength.
5. *Starch solution.*

**215.** *Absorption of Oxygen.*

(*a*) *Oxygen absorbed in four hours.*—One hundred c.c. of the water are placed in a 250 c.c. stoppered bottle and a control of 100 c.c. distilled water in a similar bottle. To each is added 10 c.c. of the permanganate solution and 10 c.c. of dilute sulphuric acid. Both bottles are then stoppered and maintained at 27° C. (80° F.) in a water-bath for four hours. At the end of that time about 10 c.c. of potassium iodide solution are added to each, whereby iodine is liberated, equivalent to the amount of potassium permanganate present, according to the equation

$$2KMnO_4 + 8H_2SO_4 + 10KI = 6K_2SO_4 + 2MnSO_4 + 8H_2O + 5I_2.$$

The iodine formed dissolves in the excess of potassium iodide and is estimated by titration with sodium thiosulphate, using starch as indicator. The blank experiment with permanganate plus distilled water serves to standardise the sodium thiosulphate, which is thus obtained direct in terms of $KMnO_4$, or available oxygen. If many analyses are to be carried out, this standardisation need not be done for each experiment, but must be repeated, nevertheless, at frequent intervals. From the amount of permanganate used up is calculated the amount of oxygen consumed per 100,000 parts of the water.

(*b*) *Oxygen absorbed in fifteen minutes.*—This test is usually only applied to effluents and very questionable waters. It is carried out in exactly the same manner as that described above, except that the samples are kept in the water-bath at 27° C. for fifteen minutes only.

Unpolluted waters show very little oxygen absorption over the shorter period, and even after four hours should not give a higher value than 0·10 per 100,000 parts of water.

## Nitrates and Nitrites.

**216.** *Estimation of Nitrites.*

*Metaphenylenediamine Method (Griess).*—This is a colorimetric estimation based on the formation of a colouring matter from metaphenylenediamine in presence of nitrites. The depth of colour is proportional to the amount of nitrites in the solution.

*Reagents.*—The following reagents are required:—

(*a*) *Standard Sodium Nitrite Solution.*—1·10 grams of pure silver nitrite are dissolved in boiling distilled water and a solution of pure sodium chloride added until no further precipitation of silver chloride occurs. The solution is cooled and made up to 1 litre with water and allowed to stand several hours. One hundred c.c. of the clear solution are then abstracted and made up to 1 litre, the dilute solution being filled into small dark glass bottles, which should be kept quite full and stored in the dark. One c.c. of this sodium nitrite solution is equivalent to 0·01 mg. of nitrogen.

(*b*) *Metaphenylenediamine solution.*—Eight grams of metaphenylenediamine hydrochloride are dissolved in water, the solution decolorised if necessary by boiling with animal charcoal, and the filtered liquid rendered slightly acid with hydrochloric acid and diluted to 1 litre with distilled water.

(*c*) *Dilute Sulphuric Acid.*—Strength 1 to 3.

*Procedure.*—Into a Nessler cylinder (as used for the ammonia estimations) are measured out 100 c.c. of the water to be tested. The outside of the cylinders should be marked at 100 c.c. content, if this has not already been done. One c.c. of standard metaphenylenediamine solution is then added, together with 1 c.c. of dilute sulphuric acid, and the liquid stirred well with a glass rod and left to stand for twenty minutes.

At the same time fractions of a cubic centimetre of standard sodium nitrite solution (0·2, 0·4, 0·6, 0·8, and 1 c.c.) are measured into separate Nessler cylinders and made up to 100 c.c. with distilled water. To each is then added 1 c.c. of metaphenylenediamine solution and 1 c.c. of dilute sulphuric acid. These solutions are allowed to stand under the same conditions as the sample test for twenty minutes, a comparison and matching of the tints being then made, as in the case of ammonia.

**217.** *Griess-Ilosvay Method.*

This method is extremely delicate, and is carried out in the same manner as the above. The solutions consist of standard sodium nitrite as before, and a mixture of (1) 0·5 gram of sulphanilic acid in 150 c.c. of dilute acetic acid, and (2) the colourless solution poured off from boiling 0·1 gram $\beta$-naphthylamine with 20 c.c. of water diluted to 150 c.c. with dilute acetic acid. An intense pink colour is rapidly produced which is proportional to the amount of nitrite present.

**218.** *Total Nitrates and Nitrites.*

The nitrates and nitrites together may be estimated in a number of ways, the nitrate figure being obtained by subtracting the value for nitrites found by one of the above methods.

(*a*) *Nitrometer Method.*—This is a very accurate method, but takes rather a considerable time to carry out. From 100 to 250 c.c. of water are evaporated to dryness on the water-bath (the residue from the "total solids" may be used), and extracted with distilled water to remove calcium and magnesium carbonates. The solution is evaporated to small bulk (about 5 c.c.) and the dissolved nitrates and nitrites estimated in the Lunge nitrometer as nitric oxide (see p. 110).

(*b*) *Estimation by Reduction to Ammonia.*—A few pieces of zinc are immersed in a dilute solution of copper sulphate until they are just covered with a firm deposit of precipitated copper. The "copper-zinc couple" thus obtained is washed with distilled water; it has the power of reducing both nitrates and nitrites with the formation of ammonia.

About 5 grams of the freshly prepared copper-zinc couple are placed in a stoppered bottle together with 100 c.c. of the water and about 1 gram of pure oxalic acid to precipitate lime. After about twelve hours an aliquot portion of the clear liquid, more or less according to the concentration of the nitrates and nitrites in the water, is transferred to a Nessler glass, diluted if necessary to 50 c.c. and the contained ammonia estimated by Nessler reagent in the ordinary way. If any free ammonia has been found in the water this must be deducted, previous to calculation of the nitrates and nitrites.

The amounts of nitrites and nitrates present in the water are calculated as parts of nitrogen per 100,000 of water.

Nearly all waters contain nitrates, generally as the product of the oxidation of nitrogenous organic matter. Nitrites are also occasionally met with, either as an intermediate stage in the oxidation, or by deoxidation of nitrates. The presence of nitrates in itself does not affect the quality of a water, but it generally points to past pollution if any quantity is present. In most cases nitrites obviously imply more recent sewage contamination. A general average content cannot very well be stated, since the conditions vary considerably in different districts and for different classes of water. In the London water supply the nitrogen as nitrate and nitrite generally averages about 0·25 part. Anything above 0·5 part is suspicious, although well-waters frequently give much higher values.

## 219. Chlorides.

The determination of chlorides is of importance in water analysis, as their presence frequently indicates sewage pollution, besides being harmful for many industrial purposes. If much magnesium chloride is present in water used for boilers, hydrochloric acid is liberated by the superheated steam and frequently causes considerable trouble.

Chlorides are estimated by titration with standard silver nitrate solution, using potassium chromate as indicator (p. 108). The silver solution used is $\frac{N}{100}$ and 1 c.c. = 0·000355 gram Cl.

One hundred c.c. of the water are measured into a porcelain basin, or into a beaker placed upon a white tile, and the silver nitrate is run in with constant stirring until a faint red tint, due to silver chromate, is observed. The volume of solution used is read off and the equivalent amount of chlorine calculated from it. A blank should be carried out with distilled water at the same time.

Water containing more than 2 parts of chlorine per 100,000 is distinctly suspicious, but its presence may, of course, be explained in some other way than by sewage pollution, and in many clay soils the uncontaminated water supply contains considerably greater proportions of chlorides.

## Acidity in Water.

Most waters are slightly alkaline towards methyl orange and sometimes also to phenolphthalein. Occasionally, however, a water will be acid to phenolphthalein, but react neutral or alkaline to methyl orange, the acidity in this case being due to dissolved carbon dioxide, towards which methyl orange is not sensitive. Many waters derived from peaty soils, are, however, acid to both indicators, the acidity being due to humic acids, etc., obtained from decaying vegetable matter. As the presence of such acids renders the water liable to dissolve the lead from pipes, its estimation in potable waters is a matter of considerable importance. It is necessary to discriminate between the acidity due to carbon dioxide and that due to other organic or inorganic acids, termed "residual acidity."

**220.** *The total acidity* is estimated in 250 c.c. of the water by titration with $\frac{N}{10}$ KOH, using phenolphthalein as indicator. The potash must be obtained from alcoholic solution, and therefore containing no carbonate.

*Residual Acidity.*—A current of air which has been first freed from carbon dioxide by passing through a

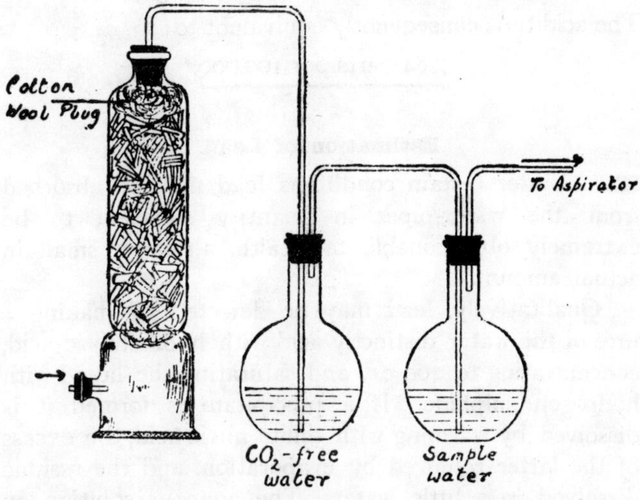


FIG. 32.—Apparatus for determining the residual acidity of water.

tower containing solid potash is drawn through 500 c.c. of the water for half an hour by means of an aspirator. The apparatus is arranged as shown in Fig 32. The acidity of 250 c.c. of the water is then determined as before, the residual acidity being thus obtained.

The acidity may be conveniently recorded in terms of the acid equivalent to the number of parts of KOH per 100,000 of water. This is obtained by multiplying the number of c.c. of $\frac{N}{10}$ KOH required for 1 litre by 0·56.

*Example:—*

$$250 \text{ c.c. required } 3{\cdot}5 \text{ c.c. } \frac{N}{10} \text{ KOH}$$

1 litre requires $3{\cdot}5 \times 4 = 14$ c.c. $= 14 \times 0{\cdot}0056$ grams KOH,

$\therefore$ 100,000 parts require $14 \times 0{\cdot}56 = 7{\cdot}84$ grams KOH.

The acidity is consequently equivalent to

7·84 parts per 100,000.

**Estimation of Lead.**

**221.** Under certain conditions lead may be absorbed from the water-pipes in quantity sufficient to be extremely objectionable to health, although small in actual amount.

Qualitatively, lead may be detected by making 1 litre of the water distinctly acid with hydrochloric acid, concentrating to 200 c.c. and saturating the liquid with hydrogen sulphide. If a precipitate is formed it is dissolved by warming with dilute nitric acid, the excess of the latter removed by evaporation, and the residue dissolved in a little water. The aqueous solution on treatment with sulphuric acid and alcohol gives a white precipitate of lead sulphate, which is again turned black by ammonium sulphide solution.

The quantitative estimation of lead is carried out colorimetrically, in absence of iron salts, as follows:—A standard lead solution containing 0·0001 gram Pb per cubic centimetre is prepared by dissolving 0·16 gram of lead nitrate in 1 litre of distilled water, or by dissolving 0·1 gram of pure lead in acetic acid and diluting to 1 litre. Four narrow cylinders of colourless glass (large Nessler glasses will do) are filled with

99, 98, 97, and 96 c.c. of distilled water, with
1, 2, 3, and 4 c.c. of the standard lead solution, equivalent to
1, 2, 3, and 4 mg. of Pb in 1 litre of mixture.

In a fifth cylinder is placed 100 c.c. of the water to be examined, together with a few drops of acetic acid. To each cylinder is now added 20 c.c. of a saturated solution of hydrogen sulphide, the liquids well mixed and the depth of colour produced in each case compared. The colour produced in the experimental water is matched with one of the standards, the depth of intermediate shades being estimated as nearly as possible. Then supposing the brown coloration to be the same as that in the cylinder with 3 c.c. of lead solution, the water contains 3 mg. of lead per litre, *i.e.* 0·3 parts per 100,000 of water.

If iron or other heavy metals are present in the water the lead must first be separated as PbS (see above), redissolved, and then estimated.

### Estimation of Hardness.

Hardness in water is generally caused by the presence of salts of calcium and magnesium, which cause the precipitation of soap. That caused by the bicarbonates is called "temporary hardness," since it can be removed by boiling the water, the normal carbonates being precipitated

$$Ca(HCO_3)_2 = CaCO_3 + CO_2 + H_2O$$

"Permanent hardness" is generally due to sulphates and chlorides and cannot be removed by boiling.

P 2

The most usual method of determining the hardness of water is to measure the amount of a soap solution required to produce a lather with the water on shaking. The soap solution is previously standardised against a solution of known strength of pure calcium chloride. Results of hardness are always stated in terms of $CaCO_3$ per 100,000 parts of water, or as grains of $CaCO_3$ per gallon.

**222.** *Estimation of Hardness by Soap Solution.*

(*a*) *Standard Soap Solution.*—Ten grams of pure Castile soap are dissolved in alcohol or methylated spirit, filtered, and made up to 1 litre with diluted spirit (about 50 per cent. by volume). A standard solution of calcium chloride is now prepared by dissolving 0·2 gram of pure crystallised calcite (Iceland spar) in dilute hydrochloric acid in a covered dish. The acid must be added gradually to prevent loss by spirting. The solution is evaporated to dryness on the water-bath, dissolved in a little distilled water, and again taken down to dryness. This operation is repeated until all the hydrochloric acid is expelled. The residue is then dissolved in pure water and the solution diluted to 1 litre.

The strong solution is standardised in the following way: 50 c.c. of the calcium chloride solution are transferred to a bottle of 250 c.c. capacity, fitted with a well ground stopper. The soap solution is run in from a burette, at first in quantities of about 2 c.c., subsequently drop by drop. After each addition the stopper is replaced and the bottle shaken vigorously six or eight times. The soap solution is run in until a soft lather is produced, which lasts without breaking for five minutes. The strong soap solution is now diluted with the aqueous spirit until exactly 14·25 c.c. are required to produce a permanent lather with 50 c.c. of the standard calcium chloride solution.

(*b*) *Total Hardness.*—Fifty c.c. of the water are placed in a stoppered bottle as above, and titrated with the soap solution in exactly the same manner described for the standardisation by calcium chloride. Care must be taken to distinguish between the real lather and the curdy scum given by magnesium salts, if present in any quantity. If the water is particularly hard and requires more than 16 c.c. of soap, it is advisable to dilute with boiled distilled water to a suitable strength and titrate an aliquot part. This is particularly necessary if the water contains much magnesium.

(*c*) *Permanent Hardness.*—To estimate the permanent hardness, 300 to 500 c.c. of the water are accurately measured into a large flask and heated to boiling for at least half an hour. It is advisable to replace part of the boiled-off water with distilled water, as otherwise the calcium sulphate itself is apt to come out of solution. After cooling to the ordinary temperature, the boiled water is run into a measuring vessel, the flask rinsed out with boiled distilled water, both water and washings being then made up to the original volume. Fifty c.c. of the filtrated water are then titrated with soap solution as before.

(*d*) *Temporary Hardness.* — This is obtained by difference from the total hardness and the permanent hardness, and represents the bicarbonates destroyed on boiling.

(*e*) *Calculation of Results.*—The volume of standard soap solution required for 50 c.c. of the water having been found, the weight of calcium carbonate corresponding to this amount in parts per 100,000 of water, may be ascertained from the following table:—

[TABLE.

### *Table of Hardness (Parts per* 100,000).

| Volume of Soap Solution. | $CaCO_3$ per 100,000. | Volume of Soap Solution. | $CaCO_3$ per 100,000. | Volume of Soap Solution. | $CaCO_3$ per 100,000. | Volume of Soap Solution. | $CaCO_3$ per 100,000. |
|---|---|---|---|---|---|---|---|
| 0·7 | 0·00 | 4·6 | 5·43 | 8·5 | 11·05 | 12·3 | 16·90 |
| 0·8 | ·16 | ·7 | ·57 | ·6 | ·20 | ·4 | 17·06 |
| 0·9 | ·32 | ·8 | ·71 | ·7 | ·35 | ·5 | ·22 |
| 1·0 | ·48 | ·9 | ·86 | ·8 | ·50 | ·6 | ·38 |
| ·1 | ·63 | 5·0 | 6·00 | ·9 | ·65 | ·7 | ·54 |
| ·2 | ·79 | ·1 | ·14 | 9·0 | ·80 | ·8 | ·70 |
| ·3 | ·95 | ·2 | ·29 | ·1 | ·95 | ·9 | ·86 |
| ·4 | 1·11 | ·3 | ·43 | ·2 | 12·11 | 13·0 | 18·02 |
| ·5 | ·27 | ·4 | ·57 | ·3 | ·26 | ·1 | ·17 |
| ·6 | ·43 | ·5 | ·71 | ·4 | ·41 | ·2 | ·33 |
| ·7 | ·56 | ·6 | ·86 | ·5 | ·56 | ·3 | ·49 |
| ·8 | ·69 | ·7 | 7·00 | ·6 | ·71 | ·4 | ·65 |
| ·9 | ·82 | ·8 | ·14 | ·7 | ·86 | ·5 | ·81 |
| 2·0 | ·95 | ·9 | ·29 | ·8 | 13·01 | ·6 | ·97 |
| ·1 | 2·08 | 6·0 | ·43 | ·9 | ·16 | ·7 | 19·13 |
| ·2 | ·21 | ·1 | ·57 | 10·0 | ·31 | ·8 | ·29 |
| ·3 | ·34 | ·2 | ·71 | ·1 | ·46 | ·9 | ·44 |
| ·4 | ·47 | ·3 | ·86 | ·2 | ·61 | 14·0 | ·60 |
| ·5 | ·60 | ·4 | 8·00 | ·3 | ·76 | ·1 | ·76 |
| ·6 | ·73 | ·5 | ·14 | ·4 | ·91 | ·2 | ·92 |
| ·7 | ·86 | ·6 | ·29 | ·5 | 14·06 | ·3 | 20·08 |
| ·8 | ·99 | ·7 | ·43 | ·6 | ·21 | ·4 | ·24 |
| ·9 | 3·12 | ·8 | ·57 | ·7 | ·37 | ·5 | ·40 |
| 3·0 | ·25 | ·9 | ·71 | ·8 | ·52 | ·6 | ·56 |
| ·1 | ·38 | 7·0 | ·86 | ·9 | ·68 | ·7 | ·71 |
| ·2 | ·51 | ·1 | 9·00 | 11·0 | ·84 | ·8 | ·87 |
| ·3 | ·64 | ·2 | ·14 | ·1 | 15·00 | ·9 | 21·03 |
| ·4 | ·77 | ·3 | ·29 | ·2 | ·16 | 15·0 | ·19 |
| ·5 | ·90 | ·4 | ·43 | ·3 | ·32 | ·1 | ·35 |
| ·6 | 4·03 | ·5 | ·57 | ·4 | ·48 | ·2 | ·51 |
| ·7 | ·16 | ·6 | ·71 | ·5 | ·63 | ·3 | ·68 |
| ·8 | ·29 | ·7 | ·86 | ·6 | ·79 | ·4 | ·85 |
| ·9 | ·43 | ·8 | 10·00 | ·7 | ·95 | ·5 | 22·02 |
| 4·0 | ·57 | ·9 | ·15 | ·8 | 16·11 | ·6 | ·18 |
| ·1 | ·71 | 8·0 | ·30 | ·9 | ·27 | ·7 | ·35 |
| ·2 | ·86 | ·1 | ·45 | 12·0 | ·43 | ·8 | ·52 |
| ·3 | 5·00 | ·2 | ·60 | ·1 | ·59 | ·9 | ·69 |
| ·4 | ·14 | ·3 | ·75 | ·2 | ·75 | 16·0 | ·86 |
| ·5 | ·29 | ·4 | ·90 | | | | |

# CHAPTER XXIII

## SOFTENING WATER FOR SPRAYS—SOFT SOAPS

IN making up sprays and washes containing soft soap it is desirable to soften water which is at all hard, in order to avoid the considerable loss of soap otherwise involved. This is done by the addition of washing-soda—$Na_2CO_3 . 10H_2O$. Permanent hardness due to sulphates, etc., is removed as follows:—

$$CaSO_4 + Na_2CO_3 = CaCO_3 + Na_2SO_4.$$

Temporary hardness, due to bicarbonates, also responds to the treatment, the bicarbonate being reduced to normal carbonate

$$Ca(HCO_3)_2 + Na_2CO_3 = CaCO_3 + 2NaHCO_3.$$

Hard waters containing much magnesium do not respond so well to softening with sodium carbonate.

**223.** (*a*) The total hardness of a water is determined as in **222** (*b*), p. 230. For each part of $CaCO_3$ per 100,000 is added to the water 2·9 parts of washing-soda and the hardness again determined.

(*b*) *Softening Water for making Soap Sprays and Washes.*—To avoid loss of expensive soap in working with a hard water, the following method is adopted:—For each "degree of hardness" (1 grain $CaCO_3$ per gallon) is added 2·86 grams of washing-soda per gallon. For each part $CaCO_3$ per 100,000 is added 2·0 grams washing-soda per gallon.

Thus, to soften 100 gallons of a hardish water of 20 degrees of hardness, one would require

$$100 \times 20 \times 2{\cdot}86 \text{ grains}$$
$$= 5720 \text{ grains} = 13 \text{ oz. washing-soda.}$$

To soften the same 100 gallons of water with soft soap of good quality, *i.e.* to produce a lather, would require $3\frac{1}{2}$-4 lbs. The relative prices and quantity required of washing-soda and soap to effect the same purpose render softening by the former very economical.

### 224. *Examination of Soft Soap.*

The examination of soft soap for making up washes, etc., is generally limited to a determination of the lathering power and the amount of carbonate present. The latter increases in quantity with the age of the soap, which should not be kept too long.

(*a*) *Lathering Power.*—It is best to compare the material with first-class soap of which the lathering power with the water of the district is known. Ten grams of the soft soap are dissolved in dilute alcohol and the solution made up to 1 litre. The amount required to produce a lather with 50 c.c. of tap-water is determined as in **222** (*a*), p. 230. The amount of soap in pounds required to give a lather with 100 gallons of the water is then calculated. Each part of soap required per 100 parts of water = one-tenth pound per gallon.

With a water of 20 degrees of hardness a good soft soap will require, to produce a lather, 3·5-4 lbs. per 100 gallons.

(*b*) *Carbonate.*—Twenty grams of the soap are dried in the steam-oven, the residue weighed and then dissolved by warming with 200 c.c. of absolute alcohol. The liquid is filtered, the undissolved residue, if any, dissolved in hot water, and the solution thus obtained titrated with $\frac{N}{10}$ sulphuric acid, using methyl orange as indicator. The alkali determined in this way is calculated as $K_2CO_3$.

# INDEX

*(The figures denote pages.)*

Acid, acetic, 16
boric, 194
citric, 47, 49
formic, 44
glyoxylic, 16
hippuric, 120
humic, 60
hydrocyanic, 44, 156
linoleic, 23
linolenic, 23
malic, 44
oleic, 23
oxalic, 45
salicylic, 195
tartaric, 46
ulmic, 60
uric, 120
Acidity in milk, 182
determination of, 182
in water, 226
determination of, 227
Acids, organic, in plant, 44
in ensilage, 169
salts of, 44
Adamkiewicz's reaction, 16
Adam's paper coil method (fat estimation), 185
Added water, in milk, detection of, 193
Albumin, in milk, estimation of, 176
Albuminoid ammonia, estimation of, 219
Albuminoids, in feeding stuffs, 150
estimation of, 150
in milk, 188
in plants, 14
Aldehydes, 50
Alkaloids, 36
occurrence of, 37, 159
reactions of, 40
Alluvial soil, analysis of, 102
Aluminium, in soil, estimation of, 100
Amides, definition of, 18
estimation of, 19, 151, 210
Ammonia, free, absorption of, by soil, 68, 69
in $H_2O$, estimation of, 218
albuminoid in $H_2O$, estimation of, 219
reagents for estimation of, in water, 216
sulphate of, 106

Ammoniacal liquor, examination of, 105
  nitrogen, in cheese, estimation of, 209
Ammonium citrate solution, 132
  compounds, absorption of, by soil, 68
    nitrification of, 65
  sulphate, absorption of, by soil, 69
    analysis of, 106
Amos' method (for carbonates), 96
Amygdalin, 42
  hydrolysis of, 43
Aniline dyes, in butter, 206
Annatto, in butter, 206
Apatite, 128
Artificial manures, 105
  nitrogenous, 105
  phosphatic, 124
  potassic, 136
Asafœtida resin, 52
Ash, in plants, estimation of, 7
    constituents of, 8
  in cheese, estimation of, 211
  in feeding stuffs, 148
Asparagine, 19

B

Bacteria, denitrifying, 66
  nitrifying, 64
    occurrence of, 66
Barley, examination of, 161
  by-products of, 165
Basic slag, 134, 143
  adulteration of, 135
  analysis of, 136
  citric solution of, 136
Becchi's test (for cotton oil), 25
Beet sugar, estimation of, 167
Bioses, 28
Biuret reaction, 15, 119
*Bixa orellana* v. Annatto, 206
Bone ash, 127
  compound, 126
  flour, 126
  meal, 127
Bones, 126
  examination of, 127
  crushed, 126
  dissolved, 126
  vitriolised, 127
Boric acid, in milk, 194
Butter, composition of, 196
  colour of, 205
  colouring matter, artificial, in, 206
    detection of, 206
  estimation of fat in, 197
    of protein in, 199
    of salt in, 198
    of solids (not fat), 198
    of water in, 196
  fat, 202
  fat of, 184
    composition of, 184, 199
    iodine value of, 204
    Reichert-Meissl No. of, 200
    saponification, equivalent of, 203
    unsaturated acids of, 204

C

Caffeine, extraction of, 37
  tests for, 42
Calcium, cyanamide, 115
  phosphates of, 125
  in soil, estimation of, 101
  nitrate, 115

Caliche, 107
Cane sugar, estimation in feeding stuffs, 153
  inversion of, 33
  specific rotation of, 32
  tests for, 28
Capillarity, measurement of, 78
Capillary attraction, in soil, 79
Carbohydrates, 26
  in feeding stuffs, estimation of, 152
  photo-synthesis of, 56
Carbonates, estimation of (by absorption), 94
    Amos' method, 96
    Schrötter's method, 57
    in soft soap, 234
Carnallit, 137
Carvacrol, 50
Casein, in cheese, 210
  in milk, 177, 190
  in vegetable, 14
Catalase, plant, 56
Cattle powders, 173
Cellulose, in plants, 26
  in shoddy, 122
  examination of, 27
  preparation of, 26
  tests for, 122
Cereals, 160
  by-products of, 165
  diastatic capacity of, 164
  examination of, 160
Cheese, analysis of, 207
  composition of, 206
  estimation of ash in, 211
    of casein in, 210
    of fat in, 208
    of nitrogen in, 209
    of phosphates in, 211
    of salt in, 211
    of water in, 207
Chloride of potash, 138
Chlorides in water, 226
Chlorophyll, preparation of, 54
Chymosin, 212
Citric acid, 47, 49
  extraction of soil by, 103
  preparation of, 47
  reactions of, 49
Clay, composition of, 62
  examination of, 61
  flocculation and deflocculation of, 81
  shrinkage of, on drying, 83
Colostrum, 175
Colouring matter, absorption of, by soil, 68
  in butter, 206
Colza oil, 22
  iodine value of, 24
Condiments, in feeding stuffs, 173
Coprolites, 128
Cotton seed oil, 22
Cotton oil, tests for, 24
  iodine value of, 24
Cotton seed cake, 158
Cream, removal of, from milk, 193
Crude fibre, estimation of, in feeding stuffs, 151
Cupric hydroxide reagent (for protein), 150
Cyanogenesis, 43, 157, 159
Cyanogenetic glucosides, 155, 159
Cymene, 50
Cytase, 55

## D

Deflocculation of clay, 81
Denitrification, measurement of, 66

Diastase, 55, 163
preparation of, 163
Diastatic capacity, estimation of, in cereals, 164
Dicalcium phosphate, 125
Digestibility of foods, 174
Digitalin, 42
Dissolved bones, 126
Distillation, in steam, apparatus for, 40

E

Elutriation, of soils, 88
Elutriator, Schone's, 89
Emulsin, 43
Ensilage *v.* Silage, 169
Enzymes, 42, 156, 199
Equalised guano, 141
Erdmann's reagent (for alkaloids), 41
Essential oils, 50
preparation of, 51
Ethereal salts, 50

F

Farmyard manure, examination of, 117
Fat, composition of, 19, 199
in butter, estimation of, 197
iodine value of, 204
in cheese, estimation of, 208
in milk, examination of, 177
estimation of, 185, 188
in plants, percentage of, 20
in feeding stuffs, extraction of, 149
Fats and oils, estimation of, Soxhlet's method, 19, 149
Feeding stuffs, composition of, 147
amides in, estimation of, 151
by-products used as, 165
carbohydrates, estimation of, 152
digestibility, estimation of, 174
fats, estimation of, 149
fibre, estimation of, 151
water, estimation of, 148
protein, estimation of, 150
concentrated forms of, 155
Fehling's solution, 28
Felspar, potash, 137
Fibre, crude, estimation of in feeding stuffs, 151
in cotton cakes, 158
in shoddy, nature of, 121
in shoddy, tests for, 122
in molasses feeds, 172
Fish guano, examination of, 142
Fleischmann's formula for milk, 188
Flocculation of clay, 81
Formaldehyde, in milk, detection of, by Schiff's reagent, 195
Formic acid, 44
Froehde's reagent (for alkaloids, 41
Fructose, tests for, 28
Furfural, estimation of, 154

G

Gamboge, resin, 52
Gelatine, 15, 51
Gerber's method, 185
Gliadin, estimation of, 161
Globulins, 15

Glucosazone, 29
Glucose, tests for, 28
Glucosides, 42
  cyanogenetic, 155, 159
Gluten, 160
  preparation of, 14
  in rice, 165
  in wheat, 160
Griess - Ilosvay reagent (for nitrites), 224
Griess' method (for nitrites), 223
Ground limestone, 144
Guaiacum resin, 52
Guano, 140
  equalised, 141
  examination of, 140
  fish, 142
  Ichaboe, 140
  Peruvian, 140
  phosphatic, 141
  uric acid from, 141

H

Halphen's test, for cotton oil, 25
Hardness in water, 229
  estimation of, 230
  methods of softening, 233
  table of, 232
Heat, absorption of, by soil, 81
Hippuric acid, 120
Hop resins, estimation of, 53
Hübl's method (for iodine values), 22
Humic acid, 60, 226
Humus, 59
  preparation of, 60
  retentive and absorptive powers of, 60
Hydrocyanic acid, tests for, 44, 156
Hydrogen peroxide, in milk, 195
Hydrolysis of amygdalin, 43
  of glucosides, 42
  of salacin, 42

I

Iceland spar, 230
Ichaboe guano, 141
Iodates, in nitrate of soda, 108
Iodine value, determination of, 22, 105, 204
    by Hübl's method, 22
    by Wijs' method, 204
  of nitrate of soda, 108
  of oils, 24
  use of, 22
Iron, in plant ash, detection of, 9
  in water, 215
  in soil, detection of, 64
    estimation of, 100
Invert sugar, 33

J

Java beans, examination of, 159

K

Kainit, analysis of, 138, 143
Kaolinite, 62
Kissling's method (for nicotine), 38
Kjeldahl's method, 10, 17, 189
  factors for, 18

Koettstorfer number, determination of, 203

L

Lactic acid, in milk, 182
Lactometer, 181
Lactose, 177
Lathering powder, 234
Lavender oil, preparation of, 51
Lead in $H_2O$, estimation of, 228
Lecithin, 5, 8
Legumin, 14, 15
Leguminous seeds, examination of, 159
Lewkowitsch's test (for cotton oil), 25
Lime, 144
  analysis of, 145
  as manure, 144
  in soil, detection of, 64
    removal of, 71
Limestone, analysis of, 57, 144
  as soil constituent, 57
Limonene, 50
Linseed cake, prussic acid in, 155, 156
  oil, iodine value of, 22
Linoleic acid, 23
Lunge's method for $NO_3$ estimation, 111
Lupine seeds, examination of, 159

M

Magnesia, in limestone, 145
  mixture, 131
Magnesium, in plant ash, detection of, 10
  in soil, 64
    estimation of, 102
Maize, examination of, 165
  by-products of, 166
  proteins of, 18
Malt, examination of, 162
Maltose, from barley, 55
Mandelin's reagent (for alkaloids), 41
Manganese, in soil, estimation of, 101
Mangolds, analysis of, 166
Manures, nitrogenous artificial, 105
    organic, 117
  phosphatic, 124
  potash, 136
  mixed, 140
    analysis of, 143
    examination of, 142
  mixing of, 143
Mauritius beans, 159
Mayer's reagent (for alkaloids), 41
Mechanical analysis of soil, 85, 88
*Micrococcus ureæ*, 120
Milk, 175
  acidity of, 182
  adulteration of, 191
    detection of, 192
  casein in, estimation of, 190
  composition of, 175
  constituents of, 176
  dry matter of, 183
  fat in, 184
    estimation of, by Adam's method, 185
    estimation of, by Gerber's method, 185
  Fleischmann's formula, 188
  preservatives in, detection of, 194

Milk, sampling of, 176
sour, analysis of, 190
Weibull's method, 191
specific gravity, 178, 181
sugar, 175
total solids of, estimation of, 183
Millon's reagent (for protein), 16
Mineral constituents, total, of soil, 98
Moisture, hygroscopic in soil, 75, 92
in butter, estimation of, 197
in cheese, estimation of, 207
in feeding stuffs, estimation of, 148
in plants, estimation of, 6
percentage of, 7
Molasses, analysis of, 171
feeds, 141
analysis of, 172
Monocalcium phosphate, 125
Monoses, 28
Morphine, tests for, 41
Murexide test, 42, 120
Muriate of potash, analysis of, 138
Mustard seed, in rape cake, 158
oil, 159
Myrosin, 159

Nessler's reagent, 216
Nicotine, estimation of, 38
tests for, 42
Nitrate of lime, 115
of soda, 107
analysis of, 108
Nitrates, absorption of, by soil, 69
Nitrates, estimation of nitrogen in, by Lunge's nitrometer method, 111
estimation of nitrogen in, by Schlossing nitrometer method, 112
in soil, detection of, 63
estimation of, 102
in water, estimation of, 223
Nitrification, 64
effect of acidity on, 66
essential conditions in, 64
Nitrites in water, estimation of, 223
*Nitrobacter*, 64
Nitrogen, estimation of Kjeldahl's method, 10, 17, 189
Lunge's method, 111
Schlossing's method, 112
in amides, 19
in cheese, 209
in milk, 189
in soil, 93
detection of, in plants, 3
in soil, 63
Nitrogenous manures, artificial, 105
organic, 117
waste products, 123
Nitrolim, analysis of, 116
Nitrometer, Lunge's, 111


O


Oats, proteins of, 18
Oil, cotton seed, examination of, 24
linseed, iodine value of, determination of, 22
Oil cakes, qualitative tests for, 155


Q

Oils, essential, 50
  estimation of, Soxhlet's method, 20
    in feeding stuffs, 149
  iodine values of, 24
  percentage of, in plants, 20
Organic acids, 44
    in silage, 169
    in silage, detection of, 170
    salts of, in plants, 44
  manures, nitrogenous, 117
  matter in soil, estimation of, 92
    in water, estimation of, 216
Opoponax resin, 52
Ossein, 127
Oxalic acid, 45
*Oxalis acetosella*, 46

P

Paracasein, 212
Pea, protein of, 18
Pentosans, estimation of, 153
Pentose, 153
Peppermint oil, preparation of, 51
Peptones, 15
Perchlorates, estimation method of potash, 99
  in nitrate of soda, 114
    estimation of, 114
Percolation, rate of, through soil, 77
Peruvian guano, examination of, 140
Phaseolunatin, 155
*Phaseolus lunatus*, 159
Phenols, 50
Phenylhydrazine, reactions, 29
Phosphates, absorption of, by soil, 69
  citric soluble, 136
  in plant ash, detection of, 9
  in soil, detection of, 62
  in cheese, estimation of, 211
  in soil, estimation of, 98, 103
  reverted, estimation of, 132
  water soluble, estimation of, 130
  precipitated, 128, 173
  insoluble, in manures, 125
  soluble, in manures, 125
  rock, 130
Phosphatic guano, 141
  manures, 124
    ingredients of, 124
Phosphorus, in plants, detection of, 4
Photo-synthesis, 56
Pinene, 50
Plants, constituents of, 14
    ultimate, 1
  ash in, determination of, 7
  moisture, determination of, 6
  nitrogen, determination of, 1
  phosphorus, determination of, 4
  sulphur, determination of, 5
Platinic chloride method of potash estimation, 138
Polarimeter, estimation of sugars, by, 31, 168
Pore space in soil, determination of, 74
Poppy seed oil, 22
  iodine value, 24
Potash, absorption of, by soil, 69
  in plant ash, detection of, 10
  in soils, detection of, 62
  estimation of, by platinic chloride method, 138

Potash, estimation of, in mixed manure, 144
by perchlorate method, 99
in soil, 99, 103
felspar, 137
manures, 136
salts of, analysis of, 139
examination of, 138
Potassium sulphate, 138
chioride, 138
Preservatives in milk, detection of, 194
Proteins, crude, estimation of, Stutzer's method, 150
digestibility of, in food-stuffs, 174
estimation of, Kjeldahl's method, 17
in butter, 199
in cheese, 209
in milk, 175
Adamkiewicz's method, 16
reactions of, Biuret's reaction, 15
Vogel's reaction, 16
xantho-proteic reaction, 15
Prussic acid, 156
Pyknometer, 179

Q

Quinine, tests for, 42

R

Raffinose, 171
Rape cake, examination of, 158
dust, 142
Reducing sugars, estimation of, in feeding stuffs, 153
Reichert-Meissl number, 200
Rennet, determination of strength of, 212
Resins, 52
in hops, determination of, 53
Rice, 164
by-products of, 166
Rochelle salt, 28
Roots, general analysis of, 166

S

Salicin, 42
Salicylic acid, in milk, 195
Saligenin, 43
Salt in butter, estimation of, 198
in cheese, 211
Salts, absorption of, by soil, 68
Sand, in soil, 60
in feeding stuffs, 148
microscopical examination of, 60
Saponification number of butter, 203
Schiffs' reagent, 195
Schlossing's method (for nitrates), 112
Schrötter's apparatus (for $CaCO_3$), 58
Schweitzer's reagent (for cellulose), 27
Sewage sludge, 142
Shoddy, analysis of, 123
examination of, 121
nature of fibre in, 122
Silk, tests for, 122
Silage, 169
examination of, 170
organic acids in, 169
Sinigrin, 159
Slaked lime, 144
Soap, soft, analysis of, 234

Soap solution, 230
  sprays, 233
Sodium, in plant ash, detection of, 10
Soils, absorption of $NH_3$, 68
    of colouring matter, 68
    of heat, 81
    of salts, 68
    of water, 76
  alluvial, 102
  alumina in, 100
  bacteria in, 64
  capillary attraction in, 79
    measurement of, 80
  carbonates in, estimation of, 57, 94, 96
  chemical analysis of, 91, 104
    by Amos' method, 96
    properties of, 64
    properties of, method of expressing, 102
  classification of, 62
  colour of, 73
  denitrification in, 66
  humus in, 59
    retentive powers of, 60
  hygroscopic $H_2O$ in, 75
  manganese in, 101
  mechanical analysis of, 85, 89
  mineral constituents of, 98
  nitrification in, 64
  percolation of water through, 77
  physical properties of, 72
  plant nutrients in, 62
  removal of lime from, 71
  retention of substance by, 67
    of water by, 76
  sampling for analysis, 83, 91
  sand in, 60
Soils, sand in, microscopical examination of, 60
  specific gravity of, 73
    apparent, 74
    true, 74
  thermometry of, 81
Soxhlet's apparatus, 21
  method of estimating fats and oils, 19, 149
Soya bean oil, 22
  iodine value of, 24
Specific gravity, of milk, 177
  estimation of, 178
  of soils, 73
    apparent, 74
    true, 74
Spices, 173
Sprays, soap, 233
Starch, 34
  action of diastase on, 55
  in cereals, 160, 165
  estimation of, 153
  percentage of, in plants, 35
  preparation of and examination of, 35
Steamed bone flour, 127
Strychnine, tests for, 42
Sucrose *v.* Cane Sugar, 153
Sugar, 28
  beet, 167
  cane, 33
    inversion of, 33
    reactions with $H_2SO_4$, 28
    reactions with Fehling's solution, 28
    reactions with phenylhydrazine, 29
    reactions with silver and cobalt nitrate, 30
    specific rotation of, 32
    tests for, 28
Sugars, examination of, 28

Sugars, estimation of, 31, 152
in milk, 177
specific rotations of, 31
pentose, 153
reducing, estimation of, in feeding stuffs, 152
Sulphate of ammonia *v.* Ammonium sulphate, 106
of potash, 138
Sulphates in ash, detection of, 10
Sulphur, in plants, detection of, 5
Sulphuric acid, in soil, estimation of, 102
Superphosphate, 128, 143
analysis of, 130
examination of, 129
of bone, 128
of lime, 129
method of expressing analysis of, 134
"reverted," 132
Surface tension, 78
Swedes, analysis of, 166

T

Tannins, 51
examination of, 52
Tartaric acid, reactions of, 46
Tartrates, 46
Terpenes, 50
Thymol, 50
Tintometer, 215
Tricalcium phosphate, 125
Tryptophane, 16
Tyrosine, 16

U

Ulmic acid, 60
Urea, examination of, 119
Urea, preparation of, from urine, 118
Uric acid, tests for, 120
in guanos, 141
Urine, constituents of, 118

V

Vitriolised bones, 127
Vogel's reaction (for proteins), 16

W

Waste products, nitrogenous, 123
Water, absorption of, by soil, 75
acidity of, 226
estimation of, 227
ammonia, free, in, estimation of, 218
albuminoid, estimation of, 219
analysis of, 215
chlorides of, 226
examination of, 214
hardness of, 229
estimation of, 230
methods of softening, 233
table of, 232
lead in, estimation of, 228
nitrates of, 223
nitrites of, 223
estimation of, by Griess method, 223
estimation of, by Griess-Ilosvay method, 224
estimation of, by nitrometer, 224